Nuclear Fission Energy - Deployment, Applications, and Beyond

Edited by Pavel Tsvetkov and Piyush Sabharwall

Published in London, United Kingdom

Nuclear Fission Energy - Deployment, Applications, and Beyond
http://dx.doi.org/10.5772/intechopen.1003407
Edited by Pavel Tsvetkov and Piyush Sabharwall

Contributors
Abigayle I. Hargreaves, Alok Kumar Jha, Amir Zacarias Mesquita, Chad L. Pope, Christopher Gundersen, Ethan Lang, Eva Barker, Jaden Palmer, Jaylyn Berry, Jolanna Witt, Leo Gannon, Leonardo Yago Zwetsch Varella, Leyton Bice, Matthew Lambert, Milton Korn, Mitchell Frasure, Nelson Snow, Nima Fathi, Salvador Munoz, Sweet Shanahan, Walewska Gouveia Silva, Welen Nunes de Lima, Xiaolei Guo

First published in London, United Kingdom, 2026 by IntechOpen
IntechOpen is the global imprint of INTECHOPEN LIMITED, registered in England and Wales, registration number: 11086078, 167-169 Great Portland Street, London, W1W 5PF, United Kingdom

For EU product safety concerns: IN TECH d.o.o., Prolaz Marije Krucifikse Kozulić 3, 51000 Rijeka, Croatia, info@intechopen.com or visit our website at intechopen.com.

British Library Cataloguing-in-Publication Data
A catalogue record for this book is available from the British Library

Nuclear Fission Energy - Deployment, Applications, and Beyond
Edited by Pavel Tsvetkov and Piyush Sabharwall
p. cm.
Print ISBN 978-0-85014-912-8
Online ISBN 978-0-85014-911-1
eBook (PDF) ISBN 978-0-85014-913-5

Meet the editors

Dr. Pavel V. Tsvetkov is a Professor and Director of the Graduate Program in Texas A&M Nuclear Engineering. Since founding the Advanced Energy Systems Laboratory, Dr. Tsvetkov has made high-impact contributions to sustainable nuclear energy. He leverages advances in artificial intelligence to innovate in instrumentation and sensing. Dr. Tsvetkov has made significant contributions to the field through 16 books in nuclear engineering, complemented by an extensive portfolio of peer-reviewed publications. His impact is evidenced by over 15 intellectual property disclosures. Dr. Tsvetkov's leadership extends through his service as Associate Editor of the ASME Journal of Nuclear Engineering and Radiation Science and Vice Chair of the ASME Technical & Strategic Advisory Board. His involvement in professional organizations, such as the American Nuclear Society (ANS) and the American Society of Mechanical Engineers (ASME), among others, highlights his scholarly contributions.

Dr. Piyush Sabharwall is a renowned staff scientist at Idaho National Laboratory (INL) with over 20 years of pioneering experience in the Nuclear Science and Technology Directorate. An expert in nuclear system design, safety, thermal-hydraulics, heat transfer, and advanced reactor technologies, he is the technical area lead for the DOE-NE's Microreactor R&D Program and the Holtec SMR-300 Advanced Reactor Demonstration Project. He spearheaded the development of the gas-cooled cartridge loop for the Versatile Test Reactor program and currently manages the Thermal Fluid Systems Methods and Analysis group, overseeing pivotal projects such as RELAP5, SAM, and Pronghorn. His visionary leadership fosters strategic partnerships, enhances INL's capabilities, and propels national energy policy. He has authored two books and over 400 peer-reviewed publications.

Contents

Preface

Welcome to *Nuclear Fission Energy – Deployment, Applications, and Beyond*. This book aims to provide an overview of nuclear fission energy, exploring its potential as a sustainable energy solution to achieve net-zero carbon objectives. As nuclear fission involves splitting atoms to release energy, it offers a promising alternative to traditional chemical processes, contributing to a balanced and resilient energy system for the future.

This book covers a broad range of topics related to nuclear fission energy, including its contemporary and future uses and deployment scenarios. It is designed to provide readers with information and considerations that showcase the potential of nuclear fission energy in a sustainable energy mix. The topics include contemporary and emerging nuclear technologies, nuclear waste, deployment strategies, supply chains, licensing, and applications.

The ultimate objective of this book is to highlight the advantages and challenges of nuclear fission energy, as well as its potential to be combined with renewable sources, such as solar and wind, to create a balanced and resilient energy system for the future.

The book is divided into three sections, each addressing different aspects of nuclear fission energy. The first section, *Fission Energy Deployment Considerations*, delves into the deployment strategies and considerations for nuclear fission energy. This section is crucial for understanding the foundational elements required to implement nuclear fission as a viable energy source successfully. It includes three chapters, each providing in-depth analysis and insights into specific areas of deployment.

The first chapter, *Thermal Analysis of Natural Uranium Oxide Fuel*, discusses the importance of predicting fuel temperature during reactor operations to ensure safety. Accurate temperature prediction is vital for maintaining the integrity of the reactor and preventing potential hazards. This chapter introduces a novel model, FCCAL, designed to estimate fuel temperatures with improved accuracy. The model takes into account the impact of fission product noble gases, particularly Xenon and Krypton, on the thermal behavior of natural uranium oxide fuel. These gases can significantly impact the thermal conductivity of the fuel, potentially leading to fuel failure within the core. By understanding these dynamics, the chapter aims to enhance the safety and efficiency of nuclear reactors.

The second chapter, *Understanding Pyroprocessing Throughput in the Context of Safeguards*, explores the reprocessing of used nuclear fuel through pyroprocessing technology. Pyroprocessing is a method that enables the recycling of nuclear fuel, thereby reducing waste and enhancing resource utilization. This chapter discusses the Resonance Absorption Densitometry for Materials Assay Security Safeguards program, which aims to develop a safeguards model for commercial viability. The program addresses measurement uncertainties and the time required for input accountancy measurements, providing insights into the conditions for plant operation under safeguard requirements. By ensuring accurate and reliable measurements, the chapter contributes to the development of secure and efficient pyroprocessing techniques.

The third chapter, *Perspective Chapter: Safe Disposal and Storage of Nuclear Waste*, provides an overview of nuclear waste management. Effective waste management is essential for the long-term sustainability of nuclear energy. This chapter addresses the origin, classification, and strategies for safe immobilization and disposal of nuclear waste. It discusses the technical and political complexities of long-term disposal, emphasizing the need for both technical innovation and public engagement to ensure the safe and permanent disposal of hazardous waste. By exploring various disposal methods and their implications, the chapter aims to provide a comprehensive understanding of nuclear waste management.

The second section, *Fission Applications*, focuses on the applications of nuclear fission energy in various industries. This section highlights the versatility and potential of nuclear fission beyond traditional power generation. It includes one chapter: *Merchant Nuclear-Powered Vessels: A New Frontier in Maritime Propulsion*. This chapter explores the potential of nuclear propulsion in the maritime industry, providing a fresh perspective on sustainable maritime transportation. Nuclear propulsion offers the advantage of reducing greenhouse gas emissions while maintaining operational efficiency, making it a compelling option for the shipping industry.

The chapter discusses modern nuclear reactor designs, fuel cycles, and operational economics. It revisits historical case studies, such as the NS Savannah, to draw insights for contemporary applications. The NS Savannah was the first nuclear-powered merchant ship, and its experience provided valuable lessons for future nuclearized vessels. The chapter also evaluates the potential of turbo-electric propulsion systems for commercial shipping, highlighting their efficiency and environmental benefits. By exploring the technical and economic aspects of nuclear propulsion, the chapter aims to demonstrate its feasibility and advantages for the maritime industry.

The third section, *Technologies beyond Fission*, explores advancements in nuclear fusion reactor technology. Nuclear fusion represents the next frontier in nuclear energy, offering the potential for virtually limitless and clean energy production. This section includes one chapter: *Advancements and Challenges in Nuclear Fusion Reactors Technology: A Comprehensive Overview*. This chapter provides an overview of nuclear fusion, discussing the fundamental concepts and techniques for energy production in fusion reactors.

Nuclear fusion involves combining atomic nuclei to release energy, mimicking the processes that power the sun. The chapter explains the processes of magnetic confinement and inertial confinement, which are the two primary methods for achieving fusion. Magnetic confinement uses powerful magnetic fields to contain the hot plasma, while inertial confinement relies on intense laser or ion beams to compress the fuel. The chapter highlights the challenges and advancements in achieving sustainable fusion energy, including the development of advanced materials, plasma control techniques, and methods for energy extraction. By providing a comprehensive overview of fusion technology, the chapter aims to inspire further research and development in this promising field.

Nuclear Fission Energy – Deployment, Applications, and Beyond presents a nuanced exploration of the benefits and challenges of nuclear fission. Through this book, we aim to showcase the advantages and challenges of nuclear fission energy and its potential to be used alongside renewable sources like solar and wind. By integrating nuclear fission with renewable energy sources, we can create a balanced and resilient energy system that meets the demands of a sustainable future. The text underscores the importance of addressing technical, regulatory, and societal challenges to unlock the full potential of nuclear energy. We hope to inspire further investigation and improvement in nuclear fuel management practices, contributing to a sustainable energy future. We aim to contribute to the advancement of nuclear energy applications.

By fostering deeper insights into reactor fuel dynamics, waste management strategies, emerging applications, and future advancements in nuclear fusion, this book serves as a valuable resource for energy professionals, policymakers, and researchers alike.

We invite readers to embark on this journey into nuclear energy, armed with knowledge that will shape conversations about sustainable power for generations to come.

Pavel V. Tsvetkov
Department of Nuclear Engineering,
Texas A&M University,
College Station, Texas, United States

Piyush Sabharwall
Nuclear Science and Technology Directorate,
Idaho National Laboratory (INL),
Idaho Falls, Idaho, United States

Section 1

Fission Energy Deployment Considerations

Chapter 1

Thermal Analysis of Natural Uranium Oxide Fuel

Alok Kumar Jha

Abstract

Predicting fuel temperature during residence in a reactor is paramount for ensuring safe operations. This chapter delves into models that facilitate the evaluation of fuel parameters, aiding in informed decision-making regarding fuel-related operational aspects, ultimately reducing the reliance on expensive experiments. Fuel temperature plays a critical role in influencing fission cross-sections and maintaining fuel integrity, which in turn affects operational safety and decisions related to the retention of fuel bundles post-power transients. Our primary focus lies on the impact of fission product noble gases (FPNGs), particularly Xenon and Krypton, on the thermal behavior of natural uranium oxide fuel. We introduce a novel model, FCCAL, designed to estimate fuel temperatures with improved accuracy. Simulations utilizing the FCCAL framework indicate significant risks of fuel failure within the core, as high concentrations of Xenon can markedly degrade thermal conductivity, resulting in elevated centerline temperatures, even in heavily irradiated low-power fuel. This study underscores the critical nature of modeling FPNG accumulation in the pellet-sheath gap, exploring its overall implications for fuel performance. Additionally, we address the role of reactivity devices in incidents of fuel failure, prompting a reexamination of overlooked aspects of fuel operations. The insights provided herein aim to inspire further investigation and improvement in nuclear fuel management practices.

Keywords: fuel temperature, PHWR, neutron diffusion equation, gap gas conductance, fuel cycle

1. Introduction

The achievement of good fuel cycle economics in both thermal and fast reactors requires the fuel element to achieve a relatively high burnup without failure. To be able to predict the fuel behavior several areas of study may be defined. One set of investigations may be along the lines of thermodynamics, mathematical modeling of fission gas behavior, fuel swelling and problems associated with oxide fuel itself. Another set of investigations may be directed at the mechanical behavior of cladding materials in reactor environments, the corrosion behavior of cladding and on solid-state physics problems such as the behavior of void in metals. All these areas of activity must be integrated to arrive at comprehensive, valid performance models for

IntechOpen

complete fuel elements. Having the ability to predict fuel temperature for efficient multiphysics of steady-state as well as transient reactor operation conditions without resorting to complex and computationally expensive calculations is a lucrative proposition driving research toward developing better models for the estimation of conductivity across nuclear fuel. Various studies [1, 2] in developing models for fuel temperature estimation bear testimony to the importance of correctly predicting fuel temperature for various fuel burnups [3, 4] and fuel power. One may desire to know the temperature distribution to evaluate the resulting thermal stresses in the fuel. In PHWR where online refueling is the norm, one may desire to know the fuel temperature to make the decision of whether to recycle and insert the fuel bundle that was previously inside the core at some location having seen some full power days of operation, thereby ascertaining the suitability of the fuel to be reinserted inside the core at some different location at different flux and hence different local power. This assumes importance in loosely coupled cores as the local power profile may not be commensurate with the global power at all times, particularly during reactivity device movement or during refueling. Fuel temperature estimation is of greater importance during operational transients in the estimation of the peak temperature reached by the fuel [5].

One of the earliest and most widely quoted experimental values for the heat transfer coefficient between UO_2 and Zircalloy-2 in a vacuum was obtained by Rapier et al. and Dean. Later, many others have proposed better models for temperature estimation. Work by researchers as recent as Maria Abramova and Faisal Raza has contributed to better estimation of fuel temperature. One important distinction in the different models is the type of fuel (e.g., Natural Uranium, Enriched Uranium, MOX etc.), which leads to different conductivities for fuel pellet with burnup, gap gas conductance, and Zircalloy sheath conductivity. CANDU-type PHWR fuel is a natural uranium 37-element fuel bundle, which contain cylindrical fuel elements with a

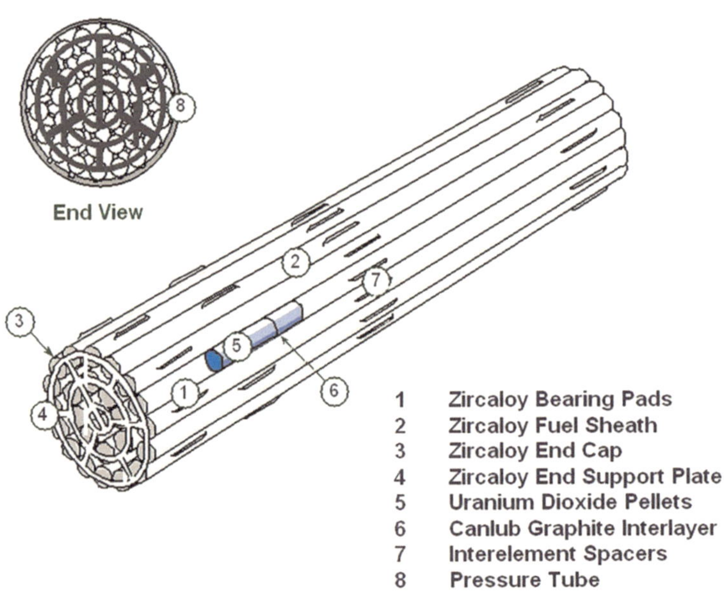

Figure 1.
CANDU-type 37-element PHWR NU fuel cluster.

13.08 mm diameter and 493 mm length, joined together by welding the elements to the end plates at both ends. The fuel bundle has one element at the center and 6, 12 and 18 elements arranged in concentric rings. Each element consists of about 32–35 sintered UO_2 pellets with a density of 10.6 Mg/m^3, 12.2 mm in diameter and 14 mm in length, enclosed in a Zircalloy-4 sheath with end plugs welded at both ends of the element [6]. A typical 37-element fuel cluster of CANDU-type PHWR NU fuel is shown in **Figure 1**.

2. Problem specification

The main factors affecting a fuel element heat transfer are geometrical configuration, neutron flux distribution, type of fuel and cladding used and coolant flow conditions. In heterogeneous reactors like PHWR, the neutrons are thermalized in the moderator outside the fuel element and absorbed in the fuel. Consequently, the neutron flux varies radially in the fuel reaching its maximum value at the fuel element periphery. Heat generation in the fuel varies in approximately the same manner. Power generation in the fuel and fuel temperature along the axis is assumed to be constant. Heat transfer is two-dimensional. In the central rod of the cluster, the heat generation is symmetrical with respect to the fuel axis. In the outer fuel rods of the cluster, an asymmetric heat source distribution prevails. Factors inherent to the fuel rod also may have an important effect on heat transfer. The heat generated in the fuel is dissipated in three ways—conduction, convection, and radiation. The heat transferred via radiation is negligible during steady-state reactor operation as the bulk of heat transfer is through conduction and convection depending upon the intervening discontinuity. Heat transfer across the fuel pencil occurs *via* three intervening media —fuel pellet (UO_2), gas in the intervening gap between the pellet and the Zircalloy sheath, the solid-solid contact points between the pellet and the Zircalloy sheath, Zircalloy sheath and the coolant film near the fuel surface. Thermal conductivity and hence fuel temperature drop across these discontinuities result in the fuel temperature profile which is nonlinear. The equivalent fuel conductivity is the sum of these components. Summarizing these effects on the heat dissipation phenomenon we may write that the heat transfer across the fuel pencil occurs along the following discontinuities.

a. Conductivity (h_P) across UO_2 pellet is

$$k = A\left[\frac{B_1}{B_2 + T} + B_3\, exp\,(B_4 T)\right] for\ 0 \le T \le 1650 \tag{1}$$

$$k = A[B_5 + B_3\, exp\,(B_4 T)]\, for\ 1650 \le T \le 2940 \tag{2}$$

where, B_1 (W/cm) = 40.4, B_2 (°C) – 464, B_3 $\left(\frac{W}{cm}{}^{\circ}C\right)$ = 1.216 × 10^{-4}, B_4 (°C^{-1}) = 1.867 × 10^{-3}, B_5(W/cm°C) = 0.0191. k is in W/cm, T is in °C and A is the porosity correction factor given by

$$A = \frac{1 - \alpha(1 - (\rho/\rho_{TD}))}{1 - \alpha(1 - 0.95)} \tag{3}$$

$$\alpha = 2.58 - 0.58 \times 10^{-3} T \tag{4}$$

b. Heat transfer across the gap and across the zircalloy surface contact points (h_{Gg} + h_{Gs} + h_s). It is estimated by the heat transfer coefficient of the pellet surface to the zircalloy outer surface in the following sub-steps.

   i. Heat transfer coefficient of the gap between the fuel pellet and the zircalloy sheath (h_{Gg}).

   ii. Heat transfer coefficient of the solid-solid contact points between the pellet and the zircalloy-4 sheath (h_{Gs}).

   iii. Heat transfer coefficient of the zircalloy-4 sheath (h_s).

Although the UO_2 pellet surface is polished to a smooth finish and the Zircalloy sheath is also planar it is a well-established fact that when nominally plane surfaces are brought together the actual area of contact is only a small fraction of the geometrical area of the interface. Even if the surfaces are carefully polished microscopic variations in the surface finish lead to contact points being limited to 'hills' of the two planar surfaces in contact. In the area wherein there is no contact between the two surfaces heat conduction by resident gas gains dominance. Irradiation of fuel induces several changes in the porosity, composition, and stoichiometry of the fuel as well as the resident gap gas composition. The heat transfer between the two surfaces in contact consists of two components—thermal conduction through solid/solid contact spots between Zircalloy sheath and UO_2 pellets, by conduction, convection, and radiation through the medium which is trapped in the voids between these spots. The solid/solid contact conductance is given by

$$h_s = \frac{k_m P}{a_0 R^{1/2} H} \tag{5}$$

where, $k_m = \frac{2k_1k_2}{k_1+k_2}$; k_m is the harmonic mean of the thermal conductivity of the two solids UO_2 ($k_1 = 0.051$ W/cm°C) and Zircalloy ($k_2 = 0.14$ W/cm°C) respectively. P (100 kg/cm^2) is the contact pressure, H (6.9×10^3 kg/cm^2) is Meyer hardness, a_0 is a constant 0.5 cm$^{1/2}$ and $R = \left(\frac{R_1^2+R_2^2}{2}\right)^{1/2}$ cm. $R_1\left(2.3 \times 10^{-4}\text{cm}\right)$ and R_2 $\left(2.3 \times 10^{-5}\text{ cm}\right)$ are arithmetic mean roughness heights of the contacting surfaces. The gap between the pellet and the pencil's inner surface is filled with Helium during manufacturing. As the fuel undergoes irradiation the fission product gases accumulating in the fuel tend to diffuse toward the outer surface of the pellet and in the gap. The dominant gases accumulating in the gap consist of fission product noble gases (Xenon and Krypton) and Iodine. Since the solubility of Xenon and Krypton in oxide fuels is very low, fuel in an operating reactor quickly becomes supersaturated in these two elements. There is a strong tendency for them to precipitate out as gas bubbles on the surrounding fuel matrix. A simple scheme showing the fraction of the fission gases accumulating in the gap is shown in **Table 1**.

From data available in the ENDF library [7], the dominant fission product gases accumulating in the gap are Xenon and Krypton along with Helium gas filled during fuel fabrication. The resultant gas accumulating in the gap has equivalent conductivity

$$\lambda_{mixture} = \sum_{i=1}^{n} \frac{\lambda_i}{1 + \sum\limits_{\substack{j=1 \\ j\neq i}}^{n} \varnothing_{ij} \frac{X_j}{X_i}} \tag{6}$$

Fraction (f)	Temperature($^{\circ}C$)
0.05	T < 1400
0.10	1500 > T > 1400
0.20	1600 > T > 1500
0.40	1700 > T > 1600
0.60	1800 > T > 1700
0.80	2000 > T > 1800
0.98	T > 2000

Table 1.
Fraction of the fission gases accumulating in the gap with temperature.

$\varnothing_{ij}$	Ternary mixture	$\varnothing_{ij}$	Binary mixture
0.219	He	0.3843	Xe
0.695	Xe	3.4284	He
0.086	Kr		

Table 2.
$\varnothing_{ij}$ of a mixture of gases accumulating in the gap.

$\varnothing_{ij}$ is a constant depending upon gas mixture composition. λ_i is the thermal conductivity coefficient of an individual gas. $\varnothing_{ij}$ for ternary gas mixture of He, Xe and Kr, and the binary gas mixture of He and Xe are given in **Table 2**.

The calculation of equivalent conductivity for ternary gas of He, Xe and Kr does not differ from the binary mixture of Xe and He by an appreciable amount. Gas gap conductivity can be assumed to be dominated by contribution from Xe accumulating from fission. The resultant equivalent conductivity of the gap between the fuel pellet and the Zircalloy sheath is $k_{eq} = h_s + \lambda_{mixture}$. Zircalloy sheath thermal conductivity is given by

$$k_{zirc}(W/mK) = 7.51 + (0.029 \times T) - \left(1.4e^{-5} \times T^2\right) + \left(7.67e^{-9} \times T^3\right) \tag{7}$$

c. Heat transfer from the pellet surface to the coolant. It is estimated by the heat transfer coefficient of the coolant film and the fuel surface (h_{Cf}).

The total heat transfer coefficient (h_T) for a fuel pencil is the sum of the terms ‘a’, ‘b’ and ‘c’ i.e.,

$$h_T = h_P + h_{Gg} + h_{Gs} + h_S + h_{Cf} \tag{8}$$

T is in Kelvin. The fuel sheath to coolant heat transfer coefficient is calculated using the Colburn correlation [8].

Eq. (9) is implemented with a code named FCCAL [9]. Fuel centerline temperature, pellet surface temperature and pencil surface temperature were calculated from the code. For the start of FCCAL iteration, a starting guess value of fuel centerline temperature is taken as 2000°C. A high value is assumed because the fuel temperature is higher than the coolant temperature for a reactor at power. Using this guess value the thermal conductivity of the pellet, gap gas, and Zircalloy sheath is computed.

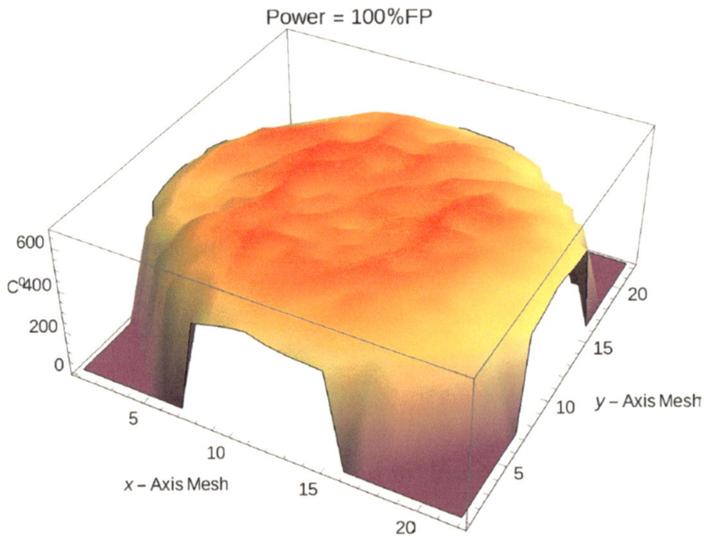

Figure 2.
Fuel temperature plot of 1830 MW PHWR core at 100%FP of operation.

Next, the temperature drops across each discontinuity is computed. The result is then compared with the benchmark 3-D neutron diffusion code solver TRIVENI [10, 11]. If the error is higher than $\pm 0.2^{\circ}$C then the simulation is repeated with a revised estimate of temperature till convergence. After convergence, the pellet centerline temperature, pencil surface temperature and coolant temperature are written to an external file for further processing. At steady reactor operation, the coolant temperature is a reliable indicator of fuel temperature. This value of steady-state temperature is input to transient calculations when the heat transfer is disrupted. In such cases the specific heat capacity of UO_2 and Zircalloy acquire significance and cannot be ignored in the estimation of the total heat load and heat removal requirement of the fuel assembly.

Fuel centerline temperature can be calculated if the power generated by the fuel bundle/pencil is known, burnup and thereby the fissile material inventory of the fuel bundle/pencil is known. The effective heat given to the coolant can be calculated using the Colburn correlation if the flow of the coolant is known. If the coolant inlet temperature is known then the resultant coolant outlet temperature can be calculated easily from the input. Burnup and flux for the corresponding fuel bundles were obtained from the 3-D finite difference solver TRIVENI. The code uses fixed axial and radial meshes for flux and power calculation of the reactor core. **Figure 2** shows the temperature profile in the reactor core. The core temperature is measured above the ambient temperature of 23°C hence the profile temperature drops down to zero at the periphery. It is notable that power generated in the core is evenly distributed with no appreciable tilt at any location.

3. Results

The model satisfactorily estimates the temperature profile to a first approximation, where the average channel power is taken as input and representative of the fuel

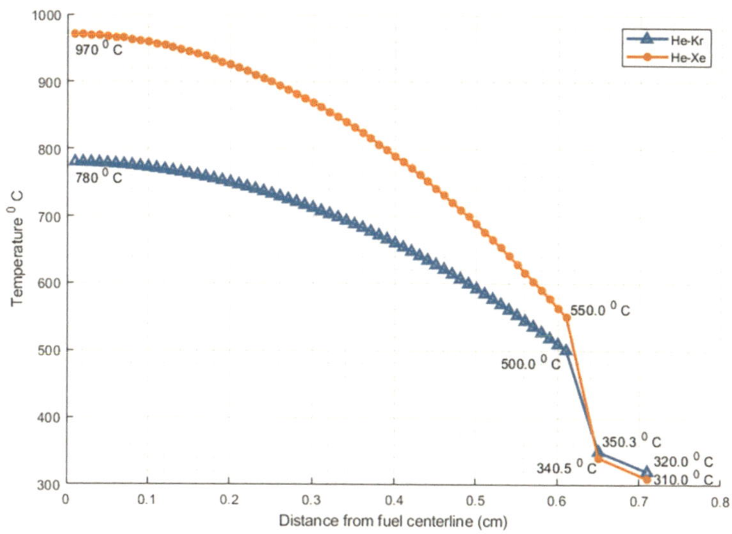

Figure 3.
Fuel temperature profile of different gas compositions but same power-producing fuel assembly.

bundle assembly residing in the reactor. The temperature profile estimated in the model gives valuable insight into the observation that there is sufficient margin to bundle power and fuel temperature in the reactor. The simulation results of low-worth centrally located devices are compared with high-worth peripherally located devices and its effect on fuel failure is presented here. The pellet centerline temperature of FCCAL shows reasonable accuracy with MATPRO [12, 13] data available in the open literature.

FCCAL result for two fuel assemblies producing similar power (600 kW) but with different gap gas compositions shows that the fuel has a different temperature profile. The result is shown in **Figure 3**.

Since the yield of Kr is small hence, He dominates the equivalent thermal conductivity of gap gas. A high yield of Xe results in a poor gap conductivity and hence higher pellet centerline temperature and higher gap temperature difference. In another study, the effect of burnup-dependent Xe accumulation in the plenum gap is presented. Two fuel assemblies are investigated which are producing similar power but different irradiation burnup. The pellet centerline temperature is higher in the high burnup fuel assembly. The effect is shown in **Figure 4**.

4. Parametric study of fuel failure

FCCAL simulation gives an indication of fuel failure in the core. It is estimated based on the following observations.

4.1 Case I: High-worth peripheral rod movement

The net reactivity worth of the peripheral rod cluster is 0.32\$. **Figure 5** shows the maximum ΔT due to rod cluster movement. The highest value is ~200°C. The greatest bundle power change is ~45 kW as shown in **Figure 6**.

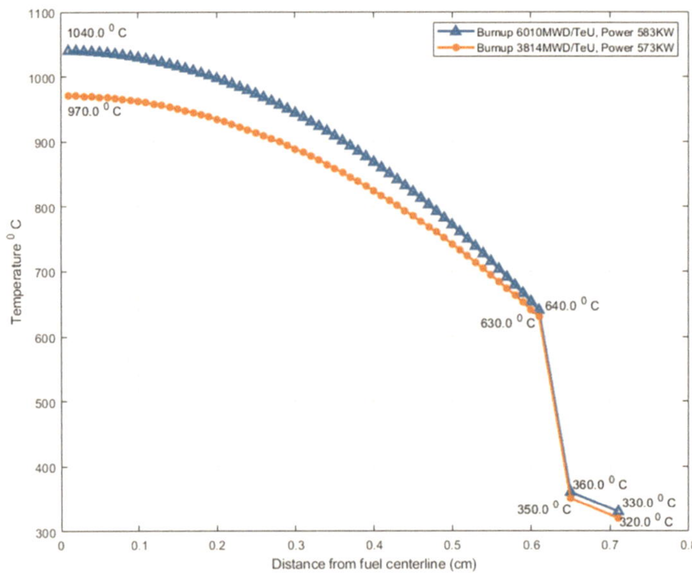

Figure 4.
Fuel temperature profile of different burnups but same power-producing fuel assembly.

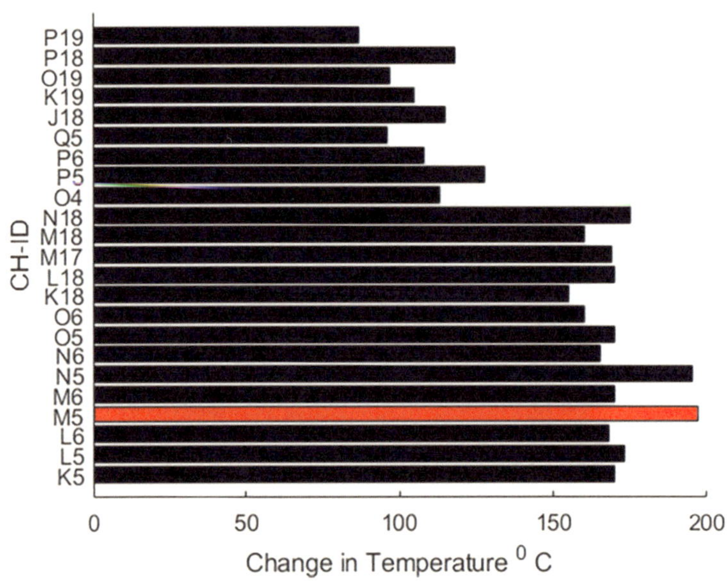

Figure 5.
Maximum ΔT due to rod cluster movement.

4.2 Case II: Low-worth central rod movement

The net reactivity worth of the peripheral rod cluster is 0.25$. **Figure 7** shows the maximum ΔT due to rod cluster movement. The highest value is ~270°C. The greatest bundle power change is ~260 kW as shown in **Figure 8**.

A comparison of Case-I and Case-II shows that the change in fuel centerline temperature and fuel power is higher for low-worth centrally located rod clusters.

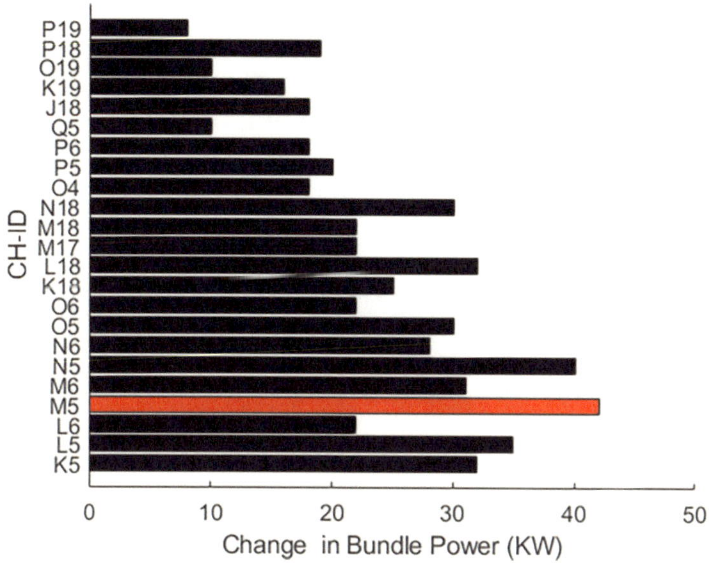

Figure 6.
Maximum bundle power change due to rod assembly movement.

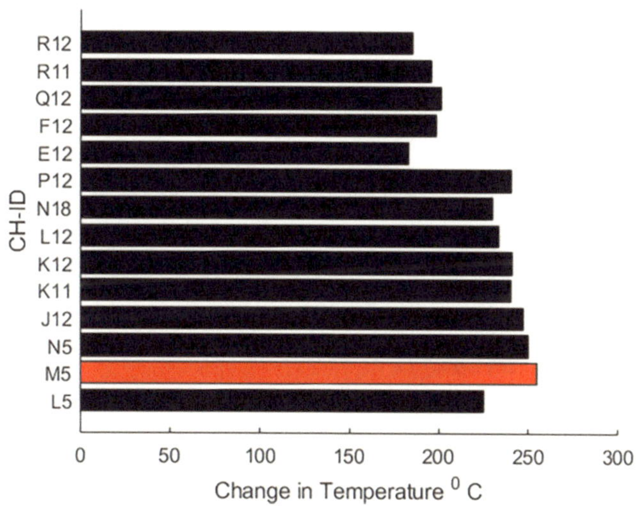

Figure 7.
Maximum ΔT due to rod cluster movement.

Operational experience shows that the central cluster movement leads to severe fuel failures whereas peripheral rod cluster movement leads to low-order failures.

In such cases, the specific heat capacity of UO_2 and Zircalloy also acquire significance and cannot be neglected in the estimation of the total heat output and heat removal requirement of the fuel assembly. Hence in that case the temperature profile estimation of the fuel pencil along with the heat capacity of UO_2 and Zircalloy gives the total heat of the fuel assembly.

The temperature of the coolant is 300°C as observed in the outlet by the instrumented channels. **Figure 5** shows the fuel centerline temperature profile of a

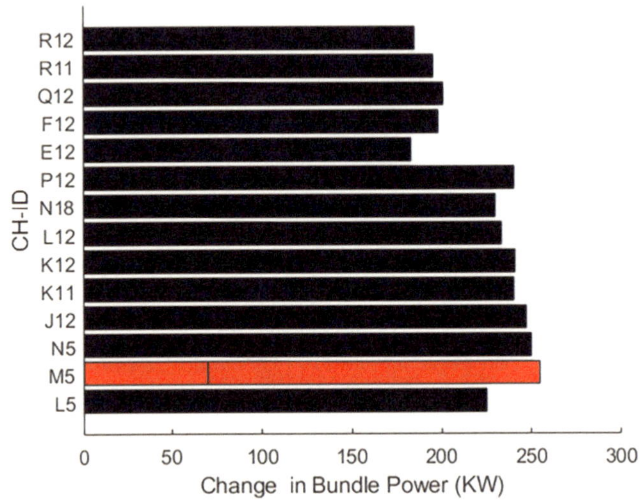

Figure 8.
Maximum bundle power change due to rod assembly movement.

peripheral channel generating low power and hence lower fuel temperature of 650°C. Lower coolant flow in the peripheral channel leads to the coolant temperature being approximately the same at 300°C.

5. Conclusion

This chapter has presented a comprehensive exploration of the fuel temperature estimation model, emphasizing the significant role of fission product noble gas (FPNG) accumulation within the pellet-sheath gap. Our findings reveal that variations in gas composition, particularly the presence of Xenon and Krypton, directly impact the thermal behavior of nuclear fuel, leading to marked differences in temperature profiles across fuel bundles.

We have established that irradiation histories contribute to unique challenges in thermal management, with gas accumulation leading to reduced thermal conductivity and elevated centerline temperatures. Moreover, the impact of reactivity device movements on operational stability has been critically assessed, highlighting potential risks associated with the current rod configuration. The observed tendency for low-worth yet centrally located devices to generate substantial power transients necessitates a reexamination of movement strategies to enhance safety margins.

The model developed in this chapter serves as a valuable tool for estimating fuel temperatures under varying operational conditions, offering insights into safe power ramps at full reactor power. Our recommendations for revising rod banking schemes aim to mitigate the risks of fuel failure, advocating for an approach that prioritizes both efficiency and safety.

In sum, this chapter underscores the importance of understanding the interplay between fuel behavior and reactor dynamics, advocating for the continuous refinement of operational practices and modeling techniques to ensure reactor safety and performance. The insights gained here lay a foundation for future research and practical applications in nuclear fuel management.

Acknowledgements

The author is indebted to Reactor Physics Section personnel of Tarapur Atomic Power Station for continuous support and fruitful discussion while preparing this manuscript.

Author details

Alok Kumar Jha
Tarapur Atomic Power Station, India

*Address all correspondence to: alokjha@npcil.co.in

References

[1] Ivanov K, Avramova M. Challenges in coupled thermal—Hydraulics and neutronics simulations for LWR safety analysis. Annals of Nuclear Energy. 2007;**34**(6):501-513

[2] Ross A, Stoute R. Heat Transfer Coefficient Between UO_2 And Zircaloy-2. Technical Report, Atomic Energy of Canada Limited; 1962

[3] Lanning D, Hann C. Review Of Methods Applicable To The Calculation Of Gap Conductance In Zircaloy-Clad UO_2 Fuel Rods. Technical Report, Battelle Pacific Northwest Labs; 1975

[4] Beyer C, Hann C, Lanning D, Panisko F GAPCON-THERMAL-2: A Computer Program for Calculating the Thermal Behavior of an Oxide Fuel Rod. 1975. Available from: https://www.osti.gov/biblio/4113463 [Accessed: April 13, 2021]

[5] Gupta S, Dutta B, Venkat R, Kokodkar A, Et Al. A Study Of The Indian PHWR Reactor Channel Under Prolonged Deteriorated Flow Conditions. Technical Report, BARC; 1997

[6] Page R Canadian power reactor fuel, Atomic Energy of Canada Ltd., Sheridan Park, Ontario: Power Projects; 1976

[7] Kinsey R. ENDF/B Summary Documentation, Technical Report. DOI: 10.2172/5405390, OSTI ID:5405390

[8] Colburn AP. A method of correlating forced convection heat-transfer data and a comparison with fluid friction. International Journal of Heat Mass Transfer. 1964;7(12):1359-1384

[9] Jha A, Gupta A, Das R, Paraswar SD. Calculation of fuel temperature profile for heavy water moderated natural uranium oxide fuel using two gas mixture conductance model for noble gas Helium and Xenon. Nuclear Engineering and Technology. 2020; **52**(12):2760-2770

[10] Srinivasan K, Purandare H. TRIVENI, 3-D Fuel Management for PHWR CANDU, Technical Report. BARC; 1985

[11] Srinivasan K. TAQUIL And TRIVENI-Computer Codes for Fuel Management of PHWRS, Note No. PHWR500/PHY/18, January, BARC; 1986

[12] Reymann GA. MATPRO Version 10: A Handbook of Materials Properties for Use in the Analysis of Light Water Reactor Fuel Rod Behavior. Idaho National Engineering Lab; 1978

[13] Feith A. Thermal Conductivity of Several Ceramic Materials to 2500 Deg C. Technical Report, General Electric Co., Cincinnati, Ohio: Nuclear Materials and Propulsion Operation; 1964

Chapter 2

Understanding Pyroprocessing Throughput in the Context of Safeguards

Eva Barker, Abigayle I. Hargreaves, Jolanna Witt, Jaylyn Berry, Leyton Bice, Mitchell Frasure, Leo Gannon, Christopher Gundersen, Matthew Lambert, Ethan Lang, Salvador Munoz, Jaden Palmer, Sweet Shanahan, Nelson Snow and Chad L. Pope

Abstract

As advanced reactors are nearing end-stage development as part of the growing effort to resolve society's energy crisis, the interest in reprocessing used nuclear fuel is increasing. Pyroprocessing is a tested reprocessing technology requiring additional study before it may be considered for commercial viability. The Resonance Absorption Densitometry for Materials Assay Security Safeguards (RADMASS) program aims to study a non-destructive assay (NDA) method and create a safeguards model that may aid in the commercial development of pyroprocessing. Measurement uncertainties and time for input accountancy measurements are incorporated into the safeguards model, which is used to inform the development of a pyroprocessing facility model. Preliminary safeguards modeling results show that the magnitude of input accountancy uncertainty is of critical importance, with other factors vary. The initial findings from the safeguards model generated measurement averages that match what is expected for a real facility. Results from modeling assist in the understanding that as throughput size of a facility increases, the standard error of inventory difference (SEID) and inventory difference (ID) values also increase. While these results represent great progress toward the commercial viability of pyroprocessing, there is still more to be done before commercialization may proceed.

Keywords: pyroprocessing, safeguards, RADMASS, modeling, uncertainty, latency, throughput

1. Introduction

In a world where electricity outages are becoming increasingly common, nuclear energy is being heavily reconsidered for the first time since the mid-twentieth century. This increased interest in nuclear energy and the innovation thereof is leading to new regulatory concerns. Fast reactors, specifically sodium-cooled fast reactors (SFRs), such as TerraPower's Natrium reactor under development, may be representative of the future of nuclear power. However, fast reactors produce used nuclear fuel (UNF) with high burnup with a considerable quantity of higher actinides.

The fuel used in SFRs will require significant planning surrounding how to manage the UNF. Uranium and transuranics (U/TRU) may be recovered through reprocessing and cycled back through a fast reactor to reduce actinide waste volumes. Reprocessing may also alleviate the cost and procurement of enough uranium to produce the higher enrichment of high-assay low-enriched uranium (HALEU) fuels needed by fast reactors, where HALEU enrichment is defined as enrichments of 5–19.95%. Reprocessing metallic fuels is compatible with a type of electrochemical separation process for UNF known as pyroprocessing.

Pyroprocessing has never been performed on a commercial scale, unlike aqueous reprocessing. It has been performed on research and engineering scales. Pyroprocessing was performed to support the Experimental Breeder Reactor II (EBR-II) in an adjacent fuel conditioning facility (FCF). In pyroprocessing, the fuel pins are chopped and put in an anode basket in a bath of molten salt called an electrorefiner (ER), where the actinides are electrically transported to collection cathodes. The uranium cathode amasses a dendrite of uranium and salt, and in further processing the salt is removed while uranium is melted into fuel pins or ingots.

A key component of any type of reprocessing is the safeguards that allow them to operate safely and with minimal proliferation risk. Safeguards are essential for the deployment of reprocessing plants to mitigate the risk of proliferation that is inherent due to the nature of the separation process. Reprocessing recovers a significant quantity of fissile material, which may be used in both nuclear weapons and for nuclear power generation. The areas of reprocessing most at risk for diversion, and therefore with the strictest safeguards, are the collection and production areas where nearly pure uranium and transuranic (U/TRU) products are found.

Safeguards vary with the regulator, though they are all intended to prevent diversion of special nuclear material (SNM) and proliferation. The International Atomic Energy Agency (IAEA), though not a regulator, defines safeguards as guidelines to aid in the verification of a State's procedure and techniques in preventing the misuse of nuclear facilities and the diversion of nuclear material from peaceful uses [1].

A guiding regulation for UNF reprocessing in the United States of America (US) is nuclear material control and accounting (MC&A), which is regulated by the US Nuclear Regulatory Commission (NRC) under 10 CFR part 74 (i.e., part 74 of title 10 of the Code of Federal Regulations) [2]. These regulations "require the licensee to maintain a nuclear [MC&A] program that tracks and verifies special nuclear material (SNM) that is on site" [2]. MC&A regulations "ensure that the information collected by the licensee about SNM is accurate, authentic, and sufficiently detailed to enable a licensee to (1) maintain current knowledge of its SNM and (2) manage its program for securing and protecting SNM" [2]. MC&A is also responsible for regulations surrounding facility record keeping, physical inventory, inventory difference (ID), and the standard error of inventory difference (SEID) [2]. Combined, these regulations aid in the safe operation of nuclear facilities in the US.

An important component of safe operation that guides UNF reprocessing is input accountancy. In aqueous reprocessing, destructive analysis (DA) is used because the fuels are dissolved in acid and can be sampled homogeneously. In pyroprocessing, the fuels are solid and nonhomogeneous. Instead, non-destructive assay (NDA) techniques must be used to determine the amount of nuclear material and isotopics with low uncertainty on the front end of the process. Additional NDA techniques are used in material balance areas (MBAs) at key measurement points (KMPs), two safeguards elements, to ensure that all materials are tracked in real time.

Established NDA methods do not provide the low uncertainty and low-latency times compatible with commercial pyroprocessing. High uncertainty or long latency in current NDA techniques prevents the ID and SEID from meeting safeguards requirements at commercial throughput paces. If ID and SEID are too high, pyroprocessing facilities will not be able to meet regulatory standards and will be unable to operate in the US.

The goal of the Resonance Absorption Densitometry for Materials Assay Security Safeguards (RADMASS) program is to aid in the modeling and determination of requirements for NDA methodology that can penetrate UNF and measure fissile mass density with <1% uncertainty with a latency of <300 seconds while being insensitive to the high background radiation that may be present inside of a reprocessing facility [3]. To meet these goals, the RADMASS team has developed a model of the dual isotope notch observer (DINO) detection system and defined parameters required of this system to obtain the uncertainty and latency characteristics described above. Furthermore, the team has also developed a safeguards facility model, the fast reactor pyroprocessing safeguards performance model (FRPSPM), and companion light water reactor pyroprocessing safeguards performance model (LWRPSPM) that are capable of taking the DINO detection modeled outputs as inputs to inform the uncertainty and latency in a hypothetical pyroprocessing facility, so that output safeguards information of inventory, ID, and SEID may be collected from the FRPSPM and investigated for compliance under many fuel conditions, throughput rates, and measurement uncertainties.

2. Pyroprocessing

Pyroprocessing begins with input accountancy for isotopics and mass characterization of a spent fuel assembly (FA). The process then diverges for oxide or metallic fuels. Metallic fuels have fewer steps. Fuel pins are then chopped inside their cladding for the placement inside anode baskets. However, oxide fuels must undergo removal of cladding to then chop or pulverize the fuel for an additional step of electrolytic reduction. The oxides are then converted into a metallic form, so that the fuel contents can dissolve in the salt bath of the ER [4]. The process for metallic and oxide fuels is identical from this point forward.

The anode baskets are placed in the ER to dissolve the fuel in chopped fuel element pieces. An ER may be either cylindrical as it is at the FCF or rectangular as designed by the Argonne National Laboratory (ANL) [5]. Within the ER, the waste and fission products are captured and separated from the remaining, viable fuel material, which is then extracted from the ER using a cathode [4].

The cathode used for pure uranium may be cylindrical, as the cathodes are in FCF, or rectangular, as designed by ANL [5]. There are also cathodes to collect transuranics, principally plutonium, with some accompanying uranium [5]. Some plutonium cathodes utilize liquid cadmium, while the ANL design is a solid, rectangular cathode [5, 6].

Most fission products remain in the salt bath once the cathode of uranium has been removed and taken to the next processing stage [4].

Upon removal of cathodes from the ER, the uranium is removed, melted, and cast into ingots, which may be shaped as either new SFR fuel pins or larger ingots and sent to storage. Both options may also involve down-blending of the final product for the sake of nonproliferation concerns [4].

A design goal for pyroprocessing is to use NDA techniques for input accountancy, with internal MC&A performed by large mass balances with high accuracy to achieve low uncertainty. DA techniques are time consuming, cost prohibitive, and require a representative homogeneous sample of materials, which is generally unavailable in pyroprocessing. The economic viability of pyroprocessing depends on the employment of high-accuracy, low-latency NDA method for input accountancy, though some DA methods may be employed in the ER, uranium product, and transuranic product to reduce error in accountancy, and verify safeguards compliance.

3. Safeguards

The advancement of new nuclear concepts with advanced reactors and metal fuels necessitates the reconsideration of pyroprocessing as a viable commercial reprocessing technique, which requires the discussion of safeguards for pyroprocessing. Safeguards approaches for reprocessing UNF are developed for commercial aqueous reprocessing. It must be assumed that pyroprocessing will be required to meet the same safeguards requirements as aqueous reprocessing in the absence of additional guidance or regulations.

No commercial pyroprocessing plant has yet been built, though the technology has been developed at experimental and engineering-scale facilities, most notably the FCF facility at the Idaho National Laboratory (INL) in the US and in the Republic of Korea (ROK). The US-based company Oklo intends to commercialize pyroprocessing and their facility is in the design phase [7]. ANL signed an agreement with Oklo in early 2022 to formalize their plans to commercialize advanced fuel recycling technology [7]. It is also notable that the State of Arkansas in the United States is currently investigating pyroprocessing as a desirable technology to be considered for the State [8].

Safeguards are overseen by the IAEA internationally and by the NRC in the US. The objective of the IAEA safeguards is “a highly probable and timely (<30 days) detection of diversion of a significant quantity of plutonium...which is taken to be 8 kg” of plutonium and deterrence of this diversion by the risk of early detection [9, 10]. The significant quantity (SQ) of low-enriched U-235 (LEU) is taken to be 75 kg [10, 11].

Both the IAEA and the NRC require calculation of the ID and SEID for a facility, known as the material unaccounted for (MUF) and σ_{mus} in IAEA terms. The ID is defined as the difference between input and output from a facility, while also considering internal gains and losses of materials according to the eq. ID = BI + A - R – EI, where BI is beginning inventory, EI is ending inventory, A is additions, and R is removals [2]. EI is a physical inventory, while BI – A + R is the book inventory [2]. Any positive or negative ID, greater than its calculated control limits, would trigger a facility’s investigation to determine the cause of the excessive ID [2].

The SEID is considered the standard deviation of the ID, which is calculated as the root of the variance for all mass measurements weighted by measurement uncertainties [2, 12]. The SEID is calculated as an evaluation of whether the ID, after a physical inventory has been done, is statistically significant [2]. SEID is also a "summation of the variances introduced by each ID measurement. ID and SEID are calculated once for a specific inventory period, to determine if material diversion may have occurred" within that inventory period and omitting contributions from common terms [11, 12].

10 CFR Part 74 contains NRC limits for the SEID, which is used to establish control limits for the ID, based on the risk significance of the material. The categories and limits are as follows:

- "Category I:
  - SEID ≥0.1% of active inventory3
  - U235 inventory difference ≥ 3 SEID and 200 grams of Pu/U233 or 300 grams of U235 high-enriched uranium (HEU)
- Category II:
  - SEID ≥0.125% of active inventory
  - U235 inventory difference ≥ 3 SEID and 200 grams of Pu/U233 or 300 grams of U235 (HEU) or 9000 grams of U235 (LEU)
- Category III:
  - Warning-level: U235 inventory difference ≥ 1.7 SEID +500 grams U235 or U inventory difference ≥ 1.7 SEID +10 kilograms of U
  - Significant inventory difference: U or U235 inventory difference ≥ 3 SEID
  - Major inventory difference: U235 inventory difference ≥ detection quantity – 1.3 SEID" [2].

The IAEA, by contrast, specifies maximum values for SEID as 2.42 kg for plutonium and 25 kg for LEU [13]. Internationally, the IAEA specifies that the SEID, or σ_{MUF} in IAEA terms, should not exceed a fraction of 1 SQ of SNM or 2.42 kg of plutonium and 25 kg of LEU [13, 14].

Safeguards approaches for reprocessing facilities have been detailed by the IAEA. It is projected that the NRC may adopt similar measures to clarify the licensing of commercial reprocessing facilities in the US. There are 10 of these traditional safeguards used at all reprocessing facilities for member, non-nuclear weapons states in accordance with the IAEA Safeguards Agreement [15]. These 10 approaches are: (1) "Material Balance Areas (MBAs) for nuclear material accounting," (2) "Defined Key Measurement Points (KMPs) for measuring the flow and inventory of nuclear material," (3) "Defined Strategic Points for containment and surveillance (C/S) and other verification measures," (4) "Nuclear Material Accountancy, supported by review of operating records and state reports," (5) "Annual Physical Inventory Verification (PIV) – typically a 'shutdown cleanout' inventory taking," (6) "Verification of domestic and international transfers of nuclear material," (7) "Statistical evaluation of the nuclear material balance to determine 'Material Unaccounted for' (MUF)," (8) "Routine, (monthly) interim inventory verifications (IIVs) for the timely detection of possible diversion of nuclear material," (9) "Verification of facility design information," and (10)" Verification of the operator's measurement system" [15]. Additional approaches have been developed for Rokkasho Reprocessing Plant, as the established methods were insufficient for a plant of its size [15]. The NRC could similarly approach safeguards on a case-by-case basis.

A commercial reprocessing facility in the US will be designated a Category II facility by the NRC, though the quantity of SNM present within the facility may suggest a Category I facility [12, 16]. NRC guidelines in 10 CFR part 74 about Material Control and Accounting (MC&A) are assumed to apply to reprocessing, in addition to regulations in 10 CFR parts 50 and 70 [2, 17, 18]. Category I facilities operate with the SEID limited to under 0.1% of active inventory (AI) and the ID is restricted to no more than 3 times the values of the SEID, with the plutonium ID no more than 200 g when ID exceeds 2 SEID [2]. Category II facilities operate with the SEID limited to under 0.125% of AI and the same ID less than 3 SEID. If the error margins in input accountancy are too high, it becomes unlikely that pyroprocessing can meet either NRC or IAEA requirements for ID and SEID.

AI has been defined by the NRC for licensed facilities as the sum of SNM in beginning inventory, ending inventory, additions to inventory, and removals from inventory after all common terms are excluded [12, 19]. Excluding all common terms in pyroprocessing leaves the AI as the sum of all materials in process at a given point in time. The AI is calculated for each MBA under the NRC [12]. The definition of MBAs is up to the operator, subject to approval from the NRC [12]. The amount of storage included in a facility may not be used to inflate the quantity of AI. Rather, the operator could make a case for keeping a certain amount of material in various processing stages to support continued plant operations in the event of equipment failure with a technical basis. The IAEA defines inventory in a facility's inventory assessments as all radioactive materials in that facility, so for pyroprocessing this would include the SNM outside the hot cell in both fuel assemblies (FAs) on site and actinide ingots removed to a storage vault on site [20]. Additionally, the quantity of inventory is not regulated under IAEA guidelines.

It is given in 10 CFR§74.53 as the responsibility to "Perform material balance tests on a lot or a batch basis, as appropriate, or monthly, whichever is sooner, and investigate any difference greater than 200 grams of plutonium or U-233 or 300 grams of U-235 that exceeds three times the estimated standard error of the inventory difference estimator" [21]. This instance designates rules for specified research activities and indicates an inclusion of both the ID exceeding 200 g of plutonium (Pu) and 3 times SEID to trigger reviews. The regulation has been confirmed that where "U235 inventory difference $\geq$ 3 SEID and 200 grams of Pu/U233 or 300 grams of U235 (HEU)" is to be understood as the 200 g Pu limit similarly applying when the ID < 3 SEID [2].

Physical inventory assessments are set for every 6 months by the NRC [16]. Physical inventory could occur more frequently when safeguard limits are exceeded. For example, the 200 g plutonium ID limit applies when the limit of ID < 3 SEID is exceeded and if both are not met, a physical inventory would take place.

4. NDA for input accountancy and DINO detection

The first step in pyroprocessing is input accountancy via an NDA method. Some methods of NDA that could be applied to input accountancy for pyroprocessing are gamma spectroscopy, neutron counting, calorimetry, microcalorimetry, passive neutron albedo reactivity/self-interrogation neutron resonance densitometry (PNAR/SINRD), and unified non-destructive assay (UNDA) [22–24]. Process monitoring relies heavily on the use of MBAs and KMPs to continually measure the amounts of Pu and U to ensure that all materials move throughout the system properly and are accounted for. A high-accuracy input accountancy is of primary importance to correctly assess the quantity of SNM going into processing.

The term "NDA" encompasses a large array of measurement techniques used to quantify the amount of nuclear material within a sample by measuring the amount of radiation emitted by the sample, or otherwise using measurements considered "non-destructive" because they do not alter the chemical or physical state of the material [25]. While NDA techniques have been used since the 1970s for safeguards applications, they have limitations [25].

In the case of total neutron counting, this technique "cannot distinguish among primary-neutron production, secondary-neutron production (induced fission), and neutron absorption" [26]. Total neutron counting may show that a fuel assembly, or the other sample, is emitting neutrons but it cannot indicate the origin of the neutrons. This technique also does not account for the neutron absorbers that may be placed in the sample, which can alter the number of neutrons emitted [26]. Delayed neutron counting is limited by not being able to "distinguish between secondary-neutron production (induced fission) and neutron absorption when used alone" [26]. It instead requires a strong neutron source to overcome the passive neutron flux from the spent FA, and it cannot distinguish between neutron absorption and production [26].

Delayed gamma spectroscopy also has its limitations, mainly that it "requires a strong neutron source to overcome the passive gamma-ray flux from the spent fuel assembly [26]. Attenuation of the gamma-rays from the inner pins makes it difficult to determine the absolute amount of fission" [26]. Passive gamma spectroscopy is limited by attenuation of the gamma rays from the inner pins that make it difficult to detect partial defects and the different attenuations of gamma rays of different energies that make it difficult to form accurate ratios [26]. Other NDA methods present similar limitations. Additionally, employing computational methods to model individual fuel assemblies and their isotopic compositions with sufficiently low uncertainty is too time consuming to offer commercial viability to a pyroprocessing facility inputting multiple FAs per day.

Within the reprocessing industry, it is understood that the existing input accountancy technology is a barrier to pyroprocessing's commercial viability due to high uncertainties in mass quantification of U-235 and Pu-239 or lengthy latency requirements for interrogation of UNF assemblies to produce lower uncertainty in those measurements [26]. Both latency time and high uncertainties render current input accountancy technologies insufficient for assaying UNF at a commercial pace. A novel NDA approach called dual isotope notch observer (DINO) detection is presented, which may aid in overcoming that limitation [27, 28].

DINO detection is expected to be superior to previously inadequate input accountancy methods available for pyroprocessing [26]. Furthermore, DINO detection is a type of NRF-based detection system. NRF-based detection systems primarily come in two types, direct and transmission detection, both of which seek to correlate a target's macroscopic resonance cross-section with the target's isotopic composition [27]. DINO detection begins with a high intensity, narrow bandwidth beam of photons tuned to the energy of nuclear resonance fluorescence (NRF) for an isotope of interest [27]. There are two isotopes of interest in pyroprocessing: U-235 and Pu-239 with NRF energies of 1.73 MeV and 2.14 MeV, respectively [29, 30].

The source for DINO detection is a laser Compton scattering (LCS)-based system [27]. LCS scatters a laser beam off a high-energy electron beam to produce a tunable, high-energy, high-intensity quasi-monoenergetic X-ray beam with a narrow energy bandwidth. This beam is tuned to the NRF energy of the isotope of interest for a rapid and accurate isotopic assay ideal for the non-destructive assay of the nuclear material [27].

The X-ray beam from the LCS source strikes the interrogation target, which absorbs photons with the NRF energy from the beam, causing fluorescence. UNF produces a high radiation field, which makes direct detection of the NRF of the interrogation target difficult. The remainder of the beam, now characterized by a notch at the NRF energy, then passes through a collimator to shield background and scattered radiation and isolate the uncollided beam, as shown in gray in **Figure 1**. The shielded X-ray beam next impinges on a witness foil composed of the resonant isotope. This foil may absorb what remains of the NRF energy in the beam, causing fluorescence. The emissions off the witness foil are detected by an energy-integrating detector. The amount of energy deposited is directly correlated with the areal density, or enrichment, of the resonant isotope in the interrogation target. The non-resonant witness foil measurement provides a reference for the expected emissions due to atomic interactions within the witness foil only. The measurement is then repeated using a witness foil composed of the same element as the resonant isotope but in which the resonant isotope is not present. The two measurements are used to produce a decision metric with a known relationship to the areal density of the resonant isotope, and the relationship is produced by a calibration procedure using targets of known composition [28, 31].

Literature suggests 3% uncertainty is achievable in 300–360 seconds for fuel containing about 3% U-235, a common initial enrichment for light water reactor (LWR) fuels [28]. However, initial replications in Monte Carlo N-Particle Transport (MCNP) 6.2 demonstrate dwell times on the order of about 7200 seconds for existing laser Compton scattering (LCS) source capabilities. An LCS source has not yet been developed that can

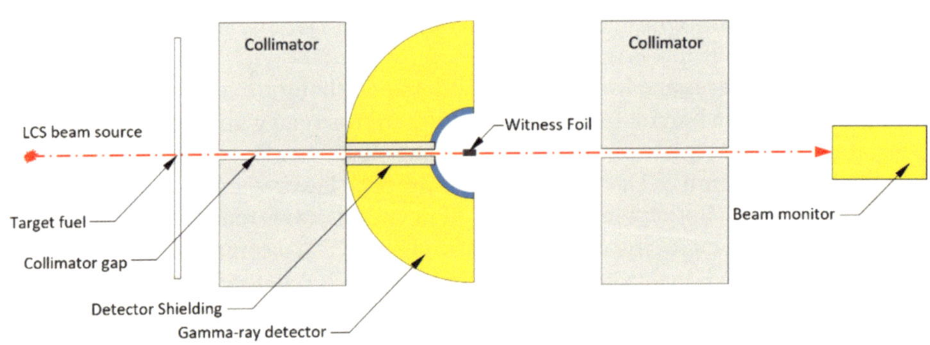

Figure 1.
DINO detection.

support a shorter dwell time by an increased flux. DINO modeling has both determined the interrogation times for various areal densities of both U-235 and Pu-239 using LCS flux values for a developed source and determined the photon fluxes needed to achieve 1% fractional standard deviation (FSD) in a 300-second dwell time.

DINO modeling shows it may be possible to interrogate an entire FA rather than single pins, depending on the strength of the LCS source. When only representative fuel pins are interrogated from each FA, the target FA will first be subject to computed tomography (CT) to select representative pins for interrogation. CT is also capable of determining if any pins have been diverted [32]. After a CT scan, the FA will be disassembled so that the selected fuel pins may be used in the DINO detection system for mass measurements of U-235 and Pu-239. Masses of other isotopes, including other plutonium isotopes, will be inferred. The bulk of the plutonium in UNF is Pu-239, however.

The RADMASS target of 1% FSD in a 300-second latency requires an LCS flux to be on the order of $10^{11}\ \gamma/s$ for U-235 and $10^{12}\ \gamma/s$ for Pu-239 in a single sodium-cooled fast reactor (SFR) fuel pin, as was determined from Monte Carlo modeling. These fluxes are beyond the capabilities of existing sources.

The interrogated and inferred contents of a FA, the time of interrogations, and accompanying error measurements with DINO may be used as inputs in the fast reactor pyroprocessing safeguards performance model (FRPSPM). In lieu of physical experimentation with DINO detection, the contents of UNF are derived from depletion analysis performed in SCALE 6.3.1 using evaluated nuclear data file version 8 (ENDF 8) for SFR fuel modeled on driver fuel fabricated for the fast flux test facility (FFTF), followed by a year of decay time. Together with the FSD error values from DINO modeling, the material content from depletion work is input to the safeguards model. The model can also inform DINO system target FSD uncertainties for U-235 and Pu-239 needed to meet safeguards requirements in SFR fuel pyroprocessing. The measurement time required to achieve that uncertainty can then be weighed against the needed time for input accountancy in the safeguards model for various throughputs.

5. Safeguards modeling

Modeling a pyroprocessing facility is a valuable method for investigating a facility's ability to meet safeguards requirements under various conditions. The goals set by the RADMASS project are first that the SEID < 0.1% of AI based on regulations from the NRC for Category I facilities, and second that the ER is not flushed out for accountancy more than once annually. Other relevant NRC regulations are that the ID does not exceed 3 times the SEID. There are two dominant models: the FRPSPM for pyroprocessing metallic fuels and the LWRPSPM for oxide fuels.

Safeguards approaches relevant to these models primarily include inventory taking using simulated mass balance measurements and the calculations of ID and SEID. Of most interest to safeguards modeling for the RADMASS project are IAEA safeguards approaches numbered 1, 2, 4, 7, and 8 [33]. The monthly interim balance verifications, which may also be called balance periods or inventory periods, are set to 30 days, as 30 days is the maximum allowable time taken to detect diversion by IAEA regulations. The NRC similarly establishes 30 days as an acceptable frequency for process monitoring [34]. MBAs, which are implied within the model, are established with KMPs using mass balance measurements. MBA mass balance measurements are also used to calculate ID and SEID for the monthly and annual inventory checks.

The FRPSPM and LWRPSPM are safeguards models developed in a Simulink subset called SimEvents. The FRPSPM is shown in **Figure 2**. The LWRPSPM is not yet optimized for safeguards studies. SimEvents enables the modeling of discrete events rather than treating all integral processes as continuous. Discrete events were chosen because they are more appropriate for the modeling of the batched and noncontinuous movements of materials occurring in pyroprocessing. The safeguards model offers a means of studying pyroprocessing throughput and mass uncertainty conditions that determine whether a pyroprocessing facility meets safeguards requirements.

In the FRPSPM, each block contains a subsystem for simulation of the movement of materials and monitoring in each process within the pyroprocessing facility [35]. These steps include the input in the UNF block in the upper left of **Figure 2**, followed by disassembly, interrogation of fuel assemblies, and then the placement of chopped fuel in anode baskets, from which point the material enters the ER. The collection of material from the ER is done with cathodes that collect either uranium (U) or uranium and transuranics (U/TRU) together, while metal waste, including cladding, is left in the anode basket, which is later collected for disposal. Outputs exit to the right-hand side of **Figure 2**. The remaining subsystems in the lower left comprise the salt cleaning and recycling system to remove the dissolved fission products and residual actinides from the salt.

The input data for the FRPSPM come from multiple sources. The material content originates in depletion analysis in SCALE 6.3.1, with the quantity of various actinides and metal groups compiled in an Excel spreadsheet, which was developed for MATLAB/Simulink compatibility. The FRPSPM incorporates the error value from the DINO detection method, and other errors, including that from in-process detectors and material balance areas. The error values chosen for input accountancy for both uranium and plutonium may be varied to identify values compatible with safeguards requirements. Other error values may also be varied. The outputs are subject to DA measurements with low errors.

The process model shown in **Figure 2** is accompanied by a monitoring system. This monitoring system simulates measurements and incorporates measurement errors, collecting data in MATLAB from the process model and outputting data by balance period for inputs, outputs, wastes, and ER inventory, and returns the amount of material in a particular process at the close of each balance period. A separate file then calculates process inventory, ID, and SEID for both U-235 and Pu using the data from the monitoring system.

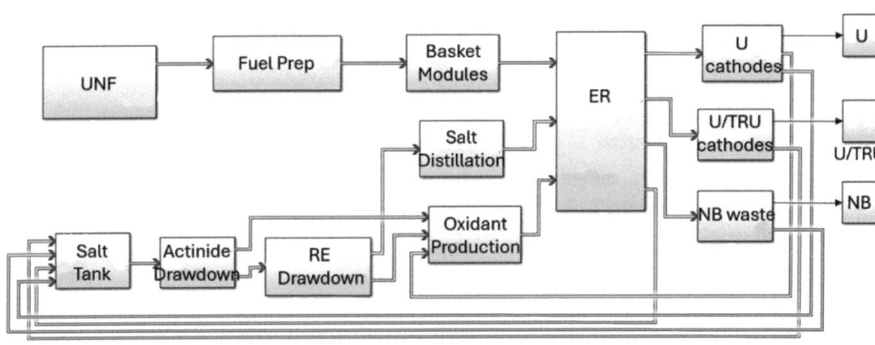

Figure 2.
Fast reactor pyroprocessing safeguards performance model.

Table 1 shows example values calculated across twelve 30-day balance periods of continual operation in the FRPSPM where the balance periods are labeled "Inv Period" and "Inv" stands for inventory. The annual throughput modeled is 50 MT annually, with fuel depletion modeled for a reactor power of 120 MW/MTU with a burnup of 20 at%, which is analogous to approximately 188 GW/MTU. Notably, the process inventory increases over each month due to the accumulation and accounting of unrecovered actinides in the ER. The contents and significance of each column will be discussed.

The inventory values reported in **Table 1** represent all materials present in processing equipment at the close of a balance period and exclude all the materials in various storage locations within the hot cell and facility. AI may include both hot cell and facility inventories depending on the MBAs designated by the operator [12]. The inventory values increase as actinides accumulate in the ER, and the values reflect the accurate accounting of that increase in material. AI is taken to be the sum of both process and hot cell storage inventories at minimum, as the MBA for pyroprocessing would not be smaller than the hot cell in which the processing takes place. If the entire plant is designated as one MBA, the SNM of the plant would all contribute to the AI, for whatever material had not been stationary for the duration of the inventory period. The regulatory differences for Category I and Category II indicate SEID < 0.1% or 0.125% of AI, respectively.

The ratio of ID to SEID values for both U-235 and plutonium is well under the ID < 3 SEID limit given by the NRC. The SEID value of plutonium exceeds the IAEA limit of 2.42 kg [13]. The LEU limit cannot be investigated here, as the total amount of uranium SEID is unreported in **Table 1**.

The Pu ID in **Table 1** is averaged to near zero, as expected for an actual facility. Errors and losses both contribute to the ID being nonzero, but the average ID should remain near zero. A negative ID value indicates that more materials exited the facility than entered and a positive ID indicates some material is left behind in the facility. When consistently positive or negative as it is displayed in the U-235 ID column of **Table 1**, it reflects an area for modeling accountancy improvement rather than actual

Inv period	U-235 Inv	U-235 ID	U-235 SEID	Pu Inv	Pu ID	Pu SEID
1	28.13	3.705	3.275	30.33	−0.253	4.303
2	28.87	3.512	3.272	31.56	−0.209	4.298
3	30.05	3.787	3.273	32.65	0.024	4.300
4	30.83	3.421	3.270	33.69	0.075	4.298
5	31.76	3.445	3.275	34.82	−0.332	4.293
6	32.57	3.776	3.273	36.00	0.270	4.301
7	33.44	3.684	3.274	37.24	−0.896	4.293
8	34.26	3.718	3.272	38.39	0.067	4.302
9	35.21	3.479	3.273	39.64	−0.172	4.300
10	36.17	3.637	3.274	40.88	−0.120	4.300
11	36.72	3.940	3.277	41.79	−0.496	4.291
12	37.61	3.750	3.271	42.80	−0.345	4.296

Table 1.
Example calculations from the FRPSPM (kg).

plant conditions. The ID for both U and Pu would be near zero with positive and negative fluctuations in a pyroprocessing facility in the absence of diversion of SNM. The calculated SEID values are constant across many modeling conditions in which ID values are affected, so the calculated SEID may be considered reliable. Furthermore, the safeguards studies underway are comparative to determine parameters with the largest impact to meeting safeguards requirements and the specific values of uranium ID are not of great significance for this work as the regulations are focused on SEID rather than ID, and the ID < 3 SEID is still true with the calculated ID values given in **Table 1**. Only when the ID exceeds 3 SEID would the regulatory limit for 200 g Pu ID take effect [34]. If this occurred, the facility would be shut down to account for the missing 200 g or more of plutonium. It is unclear whether this would necessitate the flushout of the ER. It is possible that a limited physical inventory could be approved to exclude the ER [12].

The throughput of a facility is dependent on the duration of uninterrupted operations, so the results in **Table 1** from modeling continual facility operation across 12 balance periods represent the upper bounds for the ID and SEID for any inventory period that may achieve continual operation. The maximum values of each case may be regarded as a buffer for a plant, such that ID and SEID never exceed the values calculated for continual throughput [33]. Continuous operation for extended periods of time is not practical for a facility due to safeguards compliance and plant maintenance requirements, though any commercializing organization will seek to maximize the throughput of their pyroprocessing facility.

Monthly inventory checks, annual physical inventory checks, maintenance, and equipment failures all contribute to downtime that can disrupt continuous operation and are unavoidable when considering long-term operation of a facility. Past pyroprocessing safeguards modeling and planning from ANL assumed 200 days of annual operation, while the Korean Atomic Energy Research Institute (KAERI) assumes 250 days of operation annually [5, 36]. Operations exceeding 250 days annually are unlikely to be feasible. The throughputs listed in **Table 2** equipment numbers by throughput are calculated for 200 days of operation, despite the continuous nature of the FRPSPM.

The FRPSPM has multiple operational throughput versions, as shown in **Table 2**. Equipment numbers by throughput. Different throughputs in pyroprocessing, as given by both the nominal throughput and the rate of FA processing, are supported by different numbers of internal elements, such as ERs, cathode processors (CPs), and salt drawdown vessels (DD). Process times for different FRPSPM throughputs were adjusted accordingly to reflect the number of pieces of equipment in use. Then ID, SEID, and AI are calculated using the correct number of process elements.

Throughput	FA/day	ERs	U CPs	U/TRU CPs	DD
12.5 MT	1	1	1	1	1
25 MT	2	1	2	2	2
50 MT	4	2	3	3	3
75 MT	6	3	4	5	5
100 MT	8	3	5	6	6

Table 2.
Equipment numbers by throughput.

If future SFR facilities match the generating capacity of the Fast Flux Test Facility (FFTF), three ERs are capable of processing UNF from up to 32 SFRs in a 100 MT throughput pyroprocessing facility. It is therefore unlikely an SFR pyroprocessing facility larger than 100 MT would be needed. The larger the throughput of a facility, the higher the ID and SEID values become. Alternatively, potential adopters of pyroprocessing may prefer smaller throughput models due to their ability to be used for onsite reprocessing at a SFR power plant, as was demonstrated by FCF at EBR-II.

6. Discussion and results

The first studies made using the FRPSPM explored safeguards compatibility of various fuel and measurement conditions. This study was focused on the fuel burnup, enrichment, reactor power, input accountancy error values, output and ER inventory DA error values, and throughput. Burnups were studied at 5 at%, 10 at%, 15 at%, 20 at%, and 25 at%. For reference, 1 at% of fast reactor burnup corresponds to approximately 9.4 GWd/MTHM. The reactor powers studied were 100, 110, 120, and 130 MW/MTU. U-235 enrichments tested for SFR fuels were 10%, 17.775%, and 20%, all run at 120 MW/MTU as that is the assumed power of future SFRs. FRPSPM throughputs of 12.5 MT, 25 MT, 50 MT, and 100 MT have been studied. Input accountancy and output error values were adjusted to determine the values required for safeguards compliance at the various fuel inputs, reactor powers, and model throughputs.

The target input accountancy error, or fractional standard deviation (FSD), has been determined to be constant for all throughputs, enrichments, burnups, and powers when other measurement error values are unchanged, and the AI is scaled with throughput for the limit SEID < 0.1% AI. The relationship between throughput and SEID for both U and Pu is linear and scalable, as shown in **Figure 3**. As mentioned, these SEID values are calculated under the assumption of continual processing to provide upper bounds for SEID for 30-day interim inventory assessments. In **Figure 3**, the SEID values are calculated using 1% error for input accountancy, 0.2% for the output DA values, and 0.5% for the ER and RE sampling. The AI is similarly scalable with throughput.

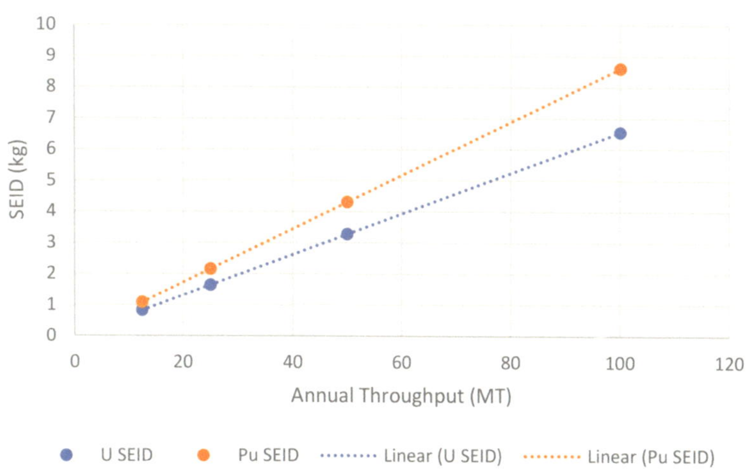

Figure 3.
SEID versus throughput for 20 at% and 120 MW/MTU.

The NRC's Pu regulation that limits ID to less than 3 times the SEID is met consistently across every burnup, throughput, enrichment, and reactor power investigated. The Pu ID regularly exceeds 200 g, but since ID <3 SEID there is no regulatory problem. The ID does not exceed 200 g for the lowest burnup of 5 at% at 25 MT throughput or less, using 30-day balance periods, for all reactor powers. However, the burnups of 5 at% and 10 at% are considered low for SFR systems and are therefore unlikely to be preferred by reactor operators.

If AI is defined for the pyroprocessing hot cell alone, this includes process inventory as well as materials stored within the hot cell awaiting the next step in processing. Storage areas include anode baskets awaiting the ER, cathodes with U and mixed U/TRU awaiting processing, and wastes containing actinides, which are batched together. The sum of all U-235 in the hot cell varies between 1000 and 7800 kg, while plutonium AI content ranges from 500 to 1900 kg depending on the burnup of the fuel. The amount of stored material in the hot cell is based on the principle that the stored material should allow the facility to continue operations for a certain time frame when a piece of equipment supplying that material breaks down. The inputs and outputs are not the same products in pyroprocessing, so the total of the inputs and outputs is also added to the storage and process inventories, for a total AI of approximately three times a 1-month of storage and process inventory sum.

If the facility inventory outside the hot cell is included in AI, it follows that AI is larger and the target input accountancy error for meeting safeguards requirements is at its maximum. In **Figure 4**, two SEID to AI variations are calculated for each throughput: one for AI for the hot cell alone and the other for the entire facility. Rather than scaling the hot cell capacity according to throughput, the quantity of material in the hot cell for each throughput is the same as 50 MT apart from some scaled material in the 100 MT model. The 50 MT model assumes that the hot cell stores enough material for 1 month of continual processing. The 100 MT model assumes a hot cell storage capacity ranging from 2 weeks to 1 month depending on the item. The 25 MT model has material sufficient for 2 months of processing in the hot cell, and the 50 MT has material for 4 months. This also means that if all throughputs had active inventories scaled for 4 months, their ratios would all have the ratios shown for 12.5 MT throughput in **Figure 4**.

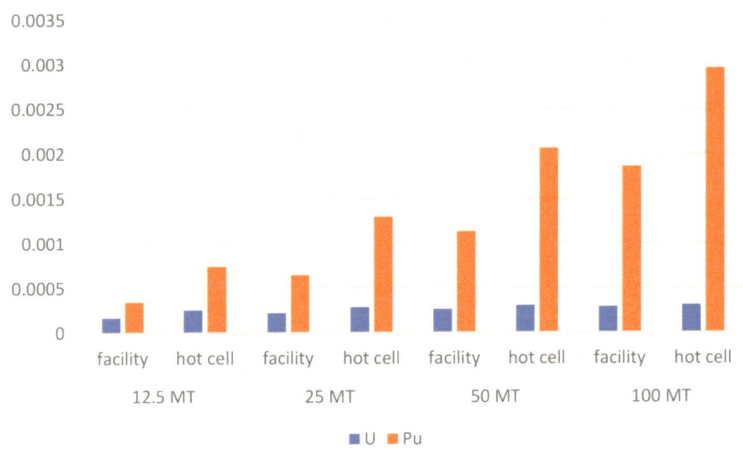

Figure 4.
SEID to active inventory ratios 20 at% and 120 MW/MTU.

For the "facility" formulations of AI, all models have material external to the hot cell equivalent to 1 month of material, as the active inventory definition necessitates material that is not counted twice and is not sitting in storage for the entirety of the balance period, which is set at one month. The facility-level material outside the hot cell in **Figure 4** includes incoming FAs, U ingots, U/TRU ingots, and noble metals (NB) waste products. The total of these is added to hot cell AI for each throughput. Given the reasoning expressed for facility AI, it may be difficult for an operator to justify storing more than 1 month of processing material in the hot cell, represented in **Figure 4** by 50 MT throughput. The horizontal line labeled 0.001 marks the point for which the RADMASS project goal and Category I limit of SEID < 0.1% AI is met.

The fuel used to produce **Figure 4** is 20 at% burnup at 120 MT/MTU reactor power. The input accountancy error values are 1%, while output error values are 0.2% for U ingots, U/TRU ingots, and noble metal waste products and 0.5% for ER and rare earth waste accountancy, taking conservative estimates from the IAEA report STR-368 (Revision 1.1), *International Target Values for Measurement Uncertainties in Safeguarding Nuclear Materials* [33, 37]. Note that the trend is the same for any fuel condition studied. Uranium meets the requirement for all throughputs and AI designations, while plutonium meets the requirement for lower throughputs using 1% input accountancy error.

Several other measurement error contributions were studied and found to have a smaller effect on the facility's ability to meet safeguards requirements when compared with input accountancy measurements for U and Pu, as shown in **Figure 5** for the case of 25 MT and 2 months of processing material in hot cell storage in AI. In changing the errors, the SEID changes, though AI is fixed. Output measurement errors, subject to destructive assay, were both doubled and halved, and the input accountancy error value was halved. Changing output error values does not lead to significantly altered SEID to AI ratios, as shown in **Figure 5** where the "Standard" is the 25 MT case from **Figure 4**. The input accountancy error values are the principal drivers in meeting safeguards requirements, as halving the input accountancy error value resulted in a significant reduction in the same SEID to AI ratio, lowering the plutonium SEID to AI ratio to meet safeguards requirements for a hot cell

Figure 5.
Error value effect on SEID to AI ratio in 25 MT throughput.

MBA and AI as well as the facility level met by the others. Therefore, 2 months of processing material in the hot cell is compatible with safeguards requirements for input error values of 0.5%.

Table 3 contains data from the study of effects of fuel enrichment on safeguards regulations. This table displays the maximum values of various parameters for any balance period, calculated for process inventory, ID, and SEID across a year of simulated data for SFR fuels with a 20 at% burnup at 120 MW/MTU, in the 50 MT throughput model, using 1% error values for both U-235 and plutonium input accountancy and process mass balance measurements, with the standard of 0.2% for output DA errors and 0.5% for ER and RE accountancy. These data demonstrate that when there is more U-235, there is less Pu, and vice versa. The magnitude of U and Pu content is irrelevant for the NRC safeguard SEID < 0.1% AI, as it involves a relative relationship rather than a set limit. As was the case for **Table 1**, the inventory columns in **Table 3** reflect only the inventory that is actively in process at the close of the balance period.

Lower burnups have more U-235 and less Pu content, as a function of known nuclear theory, while higher burnups have less U-235 and more Pu. The power in a reactor affects material content only slightly: higher power means slightly more Pu and less U-235 and vice versa. Varying enrichments for SFR fuels results in the same pattern. Even changing the throughput of a facility does not alter the required FSD if hot cell storage area volumes are scaled with the throughput. The relative importance of the burnup versus reactor power in determining the magnitude of U-235 and Pu SEID is shown in **Figure 6**, where U SEID versus burnup and power is on the left and Pu SEID versus burnup and power is on the right.

Both plots in **Figure 6** contain data calculated for a 100 MT throughput facility. Each plot shows the SEID for both burnup, using the scale on the lower x-axis, and power, using the scale on the upper x-axis. The strongly sloped lines for burnup are reported for a constant power of 120 MW/MTU, while the nearly horizontal lines for power are reported for a constant burnup of 20 at%. The difference in the two lines demonstrates that burnup drives the exchange between U-235 and Pu content far more than reactor power does. However, because the NRC safeguards for SEID < 0.1% AI or 0.125% AI are relationships rather than the fixed limit, the SEID values do not inform NRC safeguards compatibility alone.

The target input accountancy error, or fractional standard deviation (FSD), has been determined for an MBA of the hot cell alone as well as the entire facility and is the same across all burnups, enrichments, powers, and throughputs when the AI process storage is scaled to 1 month for all throughputs. The needed FSD for uranium always exceeds 1% according to the FRPSPM for all AI sizes and studies. An FSD of 1% or greater is found to meet the Category II limit of SEID <0.125% AI for the entire facility as MBA for plutonium, and 0.88% plutonium input accountancy

Enrichment	U-235 Inv	U-235 ID	U-235 SEID	Pu Inv	Pu ID	Pu SEID
10%	35.84	3.86	3.20	53.94	1.07	5.52
17.775%	36.66	3.94	3.28	42.80	0.90	4.30
20%	36.92	3.97	3.30	38.28	0.77	3.92

Table 3.
SFR fuel enrichment study using 50 MT throughput, 1% input error (kg).

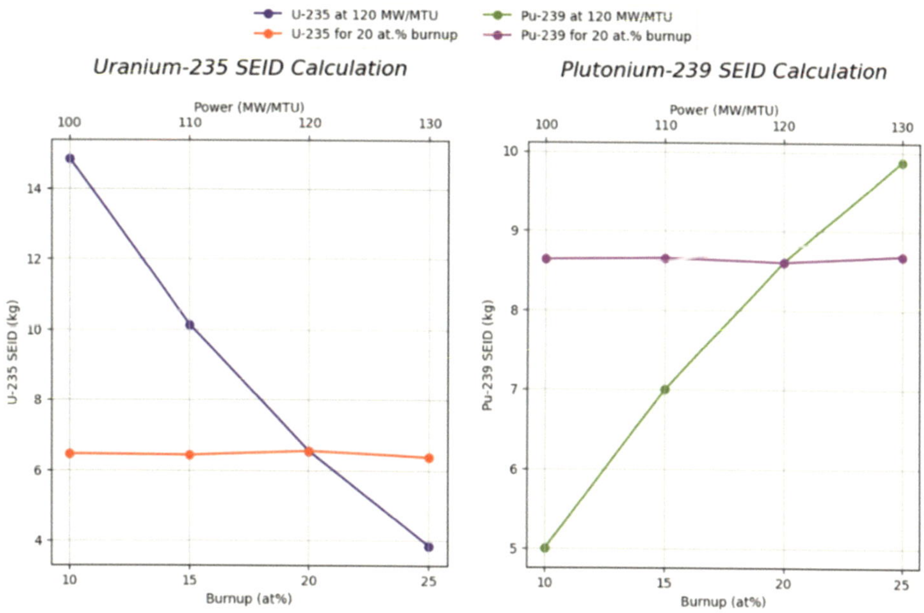

Figure 6.
SEID by burnup and reactor power (kg).

meets the Category I SEID < 0.1% AI requirement. For an MBA of the hot cell alone, the FSD needs to be 0.58% for Pu for SEID < 0.125% AI and 0.45% for Pu FSD for SEID < 0.1% AI.

The relationship between AI and input accountancy error values may be addressed in multiple ways to achieve higher throughput while maintaining low measurement uncertainty. One option is to extend the dwell time for DINO detection to lower the FSD to the desired values. The DINO detection time for a plutonium FSD of 0.45%, for a hot cell MBA under Category I, is about 9.9 hours using current LCS technology, which is too slow for commercial viability for most throughputs, except 12.5 MT, unless additional DINO detection devices are added. On the other hand, only 2.6 hours of dwell time is needed to obtain an FSD of 0.88% for Pu accountancy for a Category II hot cell MBA. In this case, 50 MT or higher could be viable commercially with additional DINO detection devices.

Another option to achieve higher throughput is to increase the storage within the hot cell, as depicted in **Figure 4**, which substantially reduces the DINO detection interrogation times because the FSD for safeguards compliance for both U-235 and Pu rises to 1% or greater. A 12.5 MT or 25 MT facility is compatible with safeguards requirements because the interrogation times do not reduce the targeted FA input pace. A 50 MT facility with 1% FSD, which requires 7200 s, or 2-hour latency under current technology, could conceivably become compatible with the utilization of an additional DINO detection device to enable the simultaneous interrogations of U-235 and plutonium, allowing the FA U-235 and Pu to both be interrogated within 6 hours and keep pace with 4 FA per day for 50 MT pyroprocessing.

The IAEA requirement for Pu is that the Pu SEID be under 2.42 kg. SEID is scalable by throughput, as shown in **Figure 3** and **Table 4**. These data show that the 25 MT throughput modeling meets the IAEA requirement for all but the highest burnup of 25 at% with the largest Pu content. For 50 MT throughput, only the lowest burnup

Throughput	U-235 SEID	LEU SEID	Pu SEID
12.5 MT	0.824	13.21	1.081
25 MT	1.641	26.29	2.156
50 MT	3.277	52.52	4.303
100 MT	6.549	104.96	8.606

Table 4.
IAEA safeguards compliance for 20 at% burnup at 120 MW/MTU (kg).

studied meets the IAEA requirement, and that burnup, 5 at%, is far less than the typical fast reactor fuel burnup. No burnups stay beneath the IAEA Pu SEID limit for 100 MT throughput.

The IAEA U SEID requirement is that the SEID be under 25 kg of LEU. Safeguards studies were performed with U-235 for NRC accountancy. The LEU SEID value was calculated using the enrichment of the UNF. **Table 4** shows IAEA safeguards compliance values for both LEU and Pu using maximum SEID values from a year of 30-day balance period calculations for the standard fuel depletion of 20 at % at 120 MW/MTU, which maintains an enrichment of 6.24 wt% in the UNF. The U SEID is scalable with throughput, as mentioned previously.

The maximum LEU SEID value for 25 MT throughput, assuming continual operation, slightly exceeds 25 kg of LEU, while the maximum Pu SEID value is slightly under the limit of 2.42 kg. Therefore, the maximum recommended throughput for pyroprocessing SFR fuels is 25 MT based on compliance with IAEA safeguards. Since a pyroprocessing facility would not operate continuously, as assumed in the modeling, it is unlikely the U limit would be exceeded under normal operation.

The RADMASS target of an ER flushout occurring no more than once annually does not seem to be an issue explicitly. Since the ER actinide content is assessed through some destructive assay measurement as modeled in the FRPSPM, there should be no cause to hold flushouts more often than annually if that accounting is demonstrated to be accurate. However, since reprocessing facilities are expected to perform 6-month physical inventory assessments, an operator would need no permission for an ER flushout exemption by demonstration of adequate accuracy in accounting through sampling or NDA methods [12].

There are many further studies in progress for both the FRPSPM and LWRPSPM. The LWRPSPM will be used to study pyroprocessing safeguards for oxide fuels with a variety of compositions, enrichments, and throughputs as has been done for the FRPSPM. Longer balance periods will be studied with both models and further statistical evaluations will be performed.

7. Conclusion

Commercializing pyroprocessing necessitates a study of safeguards requirements. Though they vary by regulatory body, each requirement designs a method of monitoring all SNMs and tracking ID and SEID to assist in the detection of any diverted material. The safeguards model presented enables a study of various conditions capable of meeting safeguards requirements.

The results of parametric studies with the FRPSPM show that there are no hurdles to reprocessing any metallic fuels via pyroprocessing due to fuel conditions.

This is supported by the initial findings from the FRPSPM that generated anticipated measurement averages that match what is expected for a real-life facility. These values from the modeling also aided in the development of the conclusion that the larger the throughput of the facility, the higher the SEID and ID values become.

However, the justification for a facility-level MBA and AI is necessary for an operator to be able to utilize an input accountancy error of near 1% rather than far less, to meet the requirement that SEID < 0.1% AI for Category I. Meeting SEID less than 0.125% of AI for Category II facilities is not problematic for input accountancy, even when considering an MBA of the hot cell alone. LWR fuels may present less of a challenge at the Category I level, and studies using the LWRPSPM are underway to determine whether this is the case.

There is an unmistakable connection between pyroprocessing facility throughput, regulatory safeguards, and commercial viability. There is still much more work and understanding that must go into pyroprocessing development. The Resonance Absorption Densitometry for Materials Assay Security Safeguards project is continuing to work towards developing an understanding of the work that must occur by both commercial and regulatory entities to make pyroprocessing a reality. This will occur via the continuation of input accountancy and modeling work as well as the development of recommended regulations for the pyroprocess.

Acknowledgements

The information, data, or the work presented herein, were funded in part by the Advanced Research Projects Agency-Energy (ARPA-E), U.S. Department of Energy (DOE), under Award Number DE-AR0001611. The views and opinions of the authors expressed herein do not necessarily state or reflect those of the United States Government or any agency thereof.

Conflict of interest

For full transparency, Evan Dolley works for GE Vernova. GE Vernova is funded by DOE/ARPA-E. ARPA-E expects publications to be made as a result of the funded program. There are no other competing interests.

Author details

Eva Barker, Abigayle I. Hargreaves, Jolanna Witt, Jaylyn Berry, Leyton Bice, Mitchell Frasure, Leo Gannon, Christopher Gundersen, Matthew Lambert, Ethan Lang, Salvador Munoz, Jaden Palmer, Sweet Shanahan, Nelson Snow and Chad L. Pope*
Idaho State University, Pocatello, Idaho, USA

*Address all correspondence to: popechad@isu.edu

References

[1] International Atomic Energy Agency. Safeguards Explained. International Atomic Energy Agency. Available from: https://www.iaea.org/topics/safeguards-explained; [Accessed: 19 July 2024]

[2] United States Nuclear Regulatory Commission. Nuclear Material Control and Accounting. 2021. Available from: https://www.nrc.gov/materials/fuel-cycle-fac/nuclear-mat-ctrl-acctng.html [Accessed: 19 July 2024]

[3] ARPA-E. Resonance Absorption Densitometry for Materials Assay Security Safeguards (RADMASS). United State Department of Energy. Available from: https://arpa-e.energy.gov/technologies/projects/resonance-absorption-densitometry-materials-assay-security-safeguards-radmass; [Accessed: 19 July 2024]

[4] Hargreaves AI, Barker E, Berry J, et al. A literature review of pyroprocessing safeguards [version 1; peer review: 3 approved with reservations]. Nuclear Science and Technology Open Research. 2024;**2**:41. DOI: 10.12688/nuclscitechnolopenres.17524.1

[5] Chang YI, Benedict RW, Bucknor MD, Figueroa J, Herceg JE, Johnson TR, et al. Conceptual Design of a Pilot-Scale Pyroprocessing Facility. Nuclear Technology. 2018;**205**(5):708-726. DOI: 10.1080/00295450.2018.1513243

[6] Willit JL, Williamson MA. High-Throughput Electrorefiner for Recovery of U and UandTRU Product from Spent Fuel. US Patent 8,097,142 B2. 2012

[7] Nuclear Engineering International. Oklo seeks licence for fuel recycling facility. Nuclear Engineering International Magazine. 2023;**27**:01

[8] Ladyman J. An Act to Create the Arkansas Nuclear Recycling Program—House Bill 1142. State of Arkansas: Arkansas State Legislature; 2023

[9]] Lineberry MJ. Personal Communication. 2014

[10] International Atomic Energy Agency. IAEA Safeguards Glossary. Vienna: International Atomic Energy Agency; 2001

[11] Riley T, Pope CL, Benedict RW. Safeguards performance model for evaluation of potential safeguards strategies applied to pyroprocessing facilities. Nuclear Engineering and Design. 2016;**301**:157-163. DOI: 10.1016/j.nucengde

[12] United States Nuclear Regulatory Commission. NUREG-2159: Acceptable Standard Format and Content for the Fundamental Nuclear Material Control Plan Required for Special Nuclear Material of Moderate Strategic Significance. Washington: Office of Nuclear Material Safety and Safeguards; 2022

[13] Chang HL, Gao FX, Ko WI, Kim HD, Lee SY. Evaluation of sigma-MUF (material unaccounted for) for the conceptually designed Korea advanced pyroprocess facility. Journal of the Korean Physical Society. 2011;**59**(2):1418-1421

[14] International Atomic Energy Agency, IAEA Safeguards Glossary. International Atomic Energy Agency; 2022

[15] Durst PC. Advanced Safeguards Approach for New Reprocessing Facilities—PNNL-16674. Pacific Northwest National Laboratory.

Richland: United States Department of Energy; 2007

[16] Nuclear Regulatory Commission. Safeguards Issues Fact Sheet: Development of Regulations for Spent Nuclear Fuel Reprocessing Facilities. Available from: https://www.nrc.gov/docs/ML1027/ML102720186.pdf

[17] United States Nuclear Regulatory Commission, Code of Federal Regulations Part 74 1985. Available from: https://www.ecfr.gov/current/title-10/chapter-I/part-74 [Accessed: 19 July 2024]

[18] United States Nuclear Regulatory Commission. Nuclear Regulatory Commission Parts 50 and 70 [NRC–2015-0016] Spent Fuel Reprocessing. United State Nuclear Regulatory Commission; 2021

[19] United States Nuclear Regulatory Commission. Instructions and Guidance for Completing Physical Inventory Summary Reports—NUREG/BR-0096, Rev 2. Office of Nuclear Material Safety and Safeguards. United States Nuclear Regulatory Commission; 2018

[20] International Atomic Energy Agency. IAEA Nuclear Security Glossary. 2020. Available from: https://www.iaea.org/sites/default/files/21/06/nuclear_security_glossary_august_2020.pdf [Accessed: 19 July 2024]

[21] United States Nuclear Regulatory Commission. 10 CFR 74.53 Process Monitoring. United States Nuclear Regulatory Commission; 2017. Available from: https://www.nrc.gov/reading-rm/doc-collections/cfr/part074/part074-0053.html [Accessed: 29 July 2024]

[22] Riley TR. Process Informed Safeguards Approach for a Pyroprocessing Facility [PhD dissertation]. Pocatello: Idaho State University; 2014

[23] Kim HD, Shin HS, Song DY, Lee TH, Ham DY, Ahn SK, et al. Application of safeguards-by-design for the pyroprocessing facilities in ROK. Journal of Nuclear Materials Management. 2012;**40**(4):24-31

[24] Coble JB, Skutnik SE, Gilliam SN, Cooper MP. Review of candidate techniques for material accountancy measurements in electrochemical separations facilities. Nuclear Technology. 2020;**206**:1803-1826

[25] Bruggeman M. Non-destructive assay of nuclear waste, 30 years of research, progress and limitations. Annalen van de Belgische Vereniging voor Stralingsbescherming. 2004;**29**(2):45-60

[26] Bolind AM, Seya M. The state of the art of the nondestructive assay of spent nuclear fuel assemblies—A critical review of the spent fuel NDA project of the U.S. Department of Energy's next generation safeguards initiative. JAEA Review. 2015;**27**:1-233

[27] Quiter BJ. Nuclear resonance fluoresence for nuclear materials assay [PhD dissertation]. Berkely: University of California; 2010

[28] Hall JM, Semenov VA, Albert F, Barty CPJ. Numerical Simulation of Nuclear Materials Detection, Imaging, and Assay with MEGa-rays. Palm Desert: INMM 52nd Annual Meeting; 2011

[29] Browne E, Tuli JK. Nuclear data sheets for a = 239. Nuclear Data Sheets. 2014;**122**:293-376

[30] Browne E, Tuli JK. Nuclear data sheets for a = 235. Nuclear Data Sheets. 2014;**122**:205-292

[31] Barty CPJ. Dual Isotope Notch Observer for Isotope Identification, Assay and Imaging with Mono-Energetic Gamma-Ray Sources. USA Patent 8,369,480. 2013

[32] Pope CL. Spent Nuclear Fuel Assembly Inspection using Neutron Computed Tomography [PhD dissertation]. Pocatello: Idaho State University; 2010

[33] Barker E, Hargreaves A, Gunderson C, Lambert M, Munoz S, Palmer J, et al. Safeguards Modeling for Studies of Pyroprocessing input Accountancy and Fuels Impact on ID and SEID. Portland: 2024 INMM Annual Meeting; 2024

[34] United States Nuclear Regulatory Commission. 10 CFR 74.53 Process Monitoring. 2017. Available from: https://www.nrc.gov/reading-rm/doc-collections/cfr/part074/part074-0053.html

[35] Barker E, Hargreaves A, Berry J. RADMASS Safeguards Report. Pocatello: Idaho State University; 2023

[36] Lee H. Basic Requirements for Preliminary Conceptual Design of the Korea Advanced Pyroprocess Facility (KAPF) - KAERI/TR-3673 /2008. Korea Atomic Energy Research Institute; 2008

[37] International Atomic Energy Agency. STR-368 (revision 1.1): International target values for measurement uncertainties in safeguarding nuclear materials. Vienna: International Atomic Energy Agency; 2022

Chapter 3

Perspective Chapter: Safe Disposal and Storage of Nuclear Waste

Xiaolei Guo

Abstract

The use of nuclear energy inevitably generates nuclear waste as the byproduct of fission reactions. Depending on the initial composition of the fuel that goes into the reactor and the subsequent burn-up level, the chemistry of the resulting nuclear waste can vary substantially. This waste typically exhibits a broad spectrum of radioactivity and half-lives, making effective management one of the most critical challenges for global nuclear energy. This chapter provides a comprehensive overview of the origin and classification of nuclear waste and various strategies for its safe immobilization and disposal. The short- and long-term storage of waste with varying radioactivity is addressed. The significant technical and political complexities involving primarily long-term disposal are also discussed. To ensure the safe and permanent disposal of hazardous waste with extremely long half-lives, future efforts should focus on both technical innovation and public engagement.

Keywords: nuclear energy, nuclear waste management, spend nuclear fuel, nuclear fuel cycle, sustainability

1. Introduction

Nuclear energy is a critical component of modern power generation, as it offers a highly efficient and low-carbon alternative to fossil fuels. The fission reaction releases an immense amount of energy, which is over a million times higher than the chemical combustion of fossil fuels [1]. The released energy is primarily manifested as heat, which is subsequently utilized to produce steam that drives turbines to generate electricity. Although the nuclear process does not emit greenhouse gases, the production of nuclear energy is inherently tied to the generation of radioactive waste, which is a byproduct that requires careful management and regulation. This waste is produced throughout the nuclear fuel cycle, from mining and enrichment to reactor operation and the decommissioning of facilities (**Figure 1**). Worldwide, nuclear power plants generate approximately 10,000 metric tonnes of heavy metal (MTHM) annually. An estimated 300,000 MTHM is currently stored globally either in pools or dry casks at reactor sites [2]. Additionally, radioactive materials used in medicine, industry, and research contribute to waste generation. Proper disposal is critical to prevent harmful radiation exposure, environmental contamination, and risks to future generations.

IntechOpen

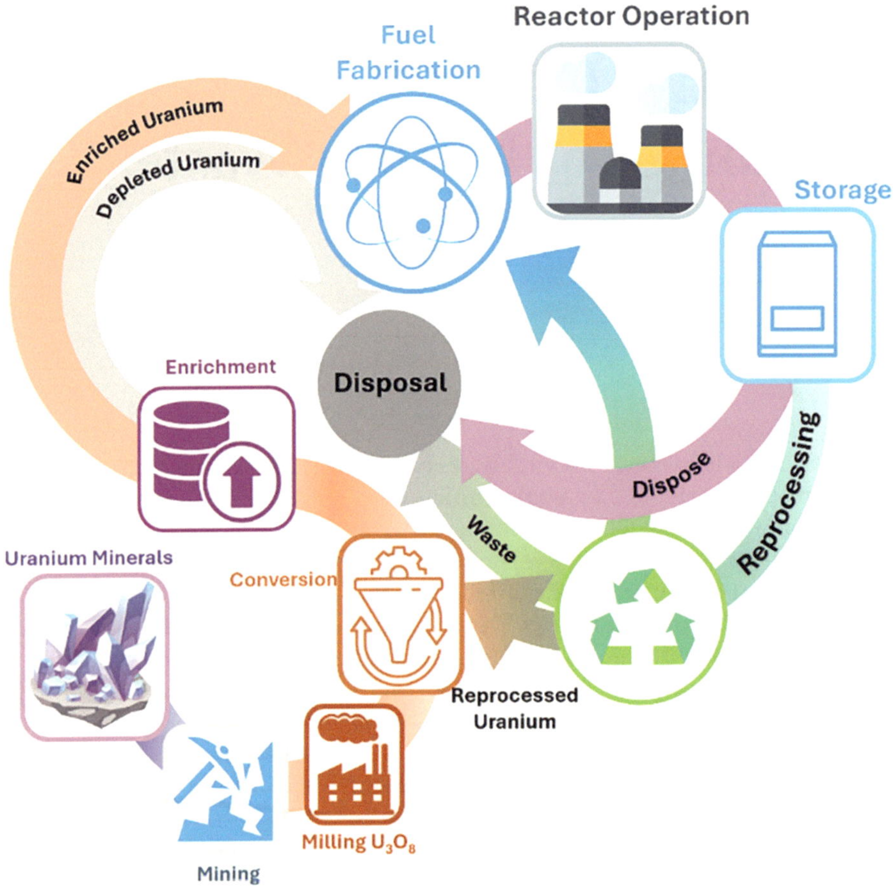

Figure 1.
Illustration of nuclear fuel cycle.

The dangers of nuclear waste stem from its radioactivity and prolonged decay periods. If nuclear waste is not properly handled, dangerous radionuclides may be released into the environment and contaminates ecosystems, potentially causing long-term health crises to human beings and irreversible environmental damage. The most challenging task to date is to ensure waste remains securely contained over geological timescales. Although waste management technologies have advanced substantially over the last several decades [3], long-term storage of radioactive waste remains a significant challenge. Therefore, continuous development of innovative solutions and effective policy frameworks remains crucial.

2. Types of nuclear waste

The chemistry and phase distribution of spent nuclear fuel (SNF) are complex and dynamic, depending primarily on its initial composition, thermal history, and burn-up level. Since the intricate chemical composition of SNF significantly affects its behavior during long-term storage and interaction with geological disposal environments, careful evaluation is essential to ensure safety over millennia. The initial state of fuel that goes into a nuclear reactor is typically in the form of solid ceramic pellets made from uranium dioxide (UO_2), which is enriched to increase the concentration of the fissile isotope ^{235}U. At the end of its reactor life, SNF still consists of about 96% UO_2, while the remaining 4% comprise fission products, transuranium elements, and activation products, distributed in a complex chemical and physical state due to the thermal and radiation environment within the reactor [1]. Fission gases such as Xe, Kr, and I are present as finely dispersed bubbles within the fuel grains, while metallic fission products like Mo, Tc, Ru, Rh, and Pd form immiscible metallic precipitates (also known as the ε-particles) ranging in size from nanometers to micrometers [4–6]. Oxide precipitates of Rb, Cs, Ba, and Zr are also present, alongside elements like Sr, Nb, and lanthanides that integrate into the UO_2 matrix as solid solutions. Additionally, transuranium elements such as ^{239}Pu substitute for U within the UO_2 lattice. The distribution of these components within a single fuel pellet is influenced by steep thermal gradients (400–1700°C within the pellet). Such steep thermal gradients can lead to grain coarsening, microfracturing, and migration of volatile elements to grain boundaries or pellet surfaces [1].

The composition and radiotoxicity of SNF evolve significantly over time due to ongoing radioactive decay and chemical changes. Initially, SNF is highly radioactive, dominated by short-lived fission products such as ^{137}Cs, ^{90}Sr, and ^{131}I, which contribute to its intense heat and radiotoxicity [1]. These isotopes decay within weeks to decades, leading to reduced immediate hazard. Over longer periods, the radiotoxicity shifts to being dominated by long-lived actinides such as ^{239}Pu, ^{241}Am, and ^{237}Np, which persist for thousands to millions of years. As the chemistry of the fuel evolves, the complexity of fission product and activation-product interactions diminishes, leaving the long-term chemistry predominantly dominated by actinides, with U making up over 95 at. %. This shift is crucial when developing strategies for the long-term containment of radionuclides [1].

Based on its radioactivity and origin, the SNF can be broadly classified into three principal categories as described below.

2.1 Low-level waste

Low-level waste (LLW) encompasses materials that have been exposed to radioactivity but contain relatively low levels of radioactive contamination compared to intermediate- and high-level waste. It is the most prevalent category of nuclear waste, originating from diverse sources such as nuclear power plants, medical facilities, industrial applications, academic research, and the decommissioning of nuclear facilities. Common examples of LLW include contaminated clothing, tools, filters, laboratory supplies, and other items that have been in contact with radioactive materials during routine operations or research.

LLW has relatively low radiation levels, so it does not require shielding during handling and transportation. Unlike high-level waste (HLW), which will be discussed below, it does not generate significant heat and contains radionuclides with shorter half-lives. These features make the management of LLW comparatively less complex. However, the proper handling and disposal of LLW are still essential to ensuring public and environmental safety. The primary issue with LLW is its high volume relative to its lower radioactivity. Therefore, disposal of LLW is often done in near-surface repositories (**Figure 2**), which is carefully managed to prevent leakage into soil or groundwater.

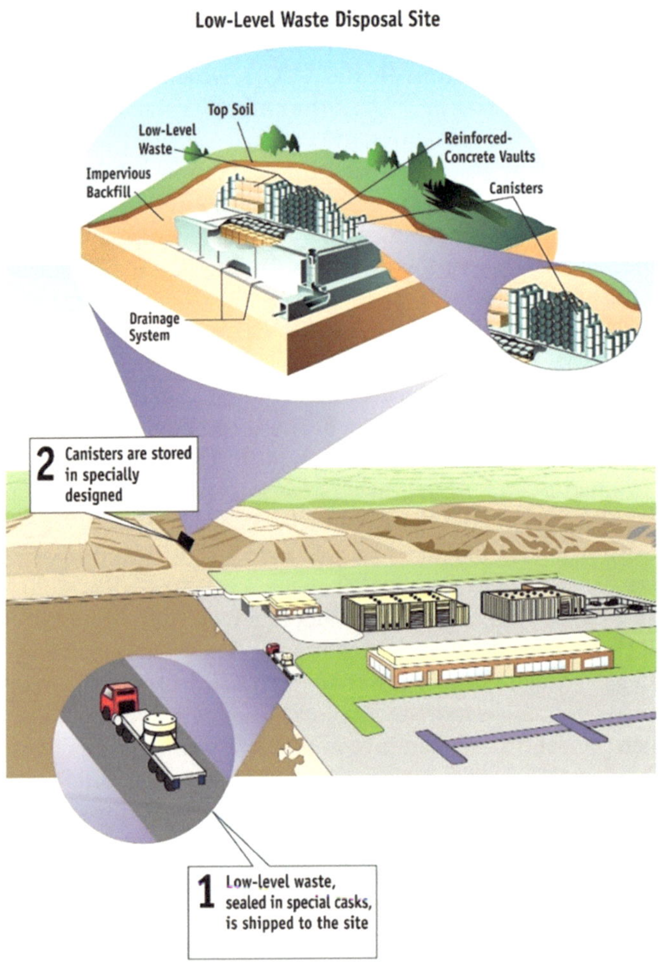

Figure 2.
Illustration of near surface disposal site for low-level waste. Source: U.S. Nuclear Regulatory Commission.

2.2 Intermediate-level waste

Intermediate-level waste (ILW) is a category of radioactive waste with higher levels of radioactivity than LLW but lower than HLW. Unlike LLW, ILW often requires shielding during handling and transportation to protect against radiation exposure due to its elevated radioactivity. Although ILW does not generate significant heat, it contains long-lived radionuclides. Therefore, careful management is needed to ensure long-term safety.

ILW is predominantly generated from the operation and maintenance of nuclear reactors, as well as the decommissioning of nuclear facilities. Typical examples of ILW include reactor components (e.g., control rods, structural materials, and pressure vessel parts), which are exposed to intense radiation during reactor operation. ILW also comes from resins and sludges from water treatment systems, as well as contaminated equipment used in reprocessing spent fuel. ILW may also arise from industrial and research applications involving radioactive materials. These diverse sources significantly enhance the complexity of the ILW.

Containment and stabilization are essential to prevent the release of radioactive materials from the ILW. This waste is typically treated, solidified, and placed in specialized containers designed to withstand the effects of radioactivity over extended periods. These containers are then stored in tailored facilities, often located deep underground or in specially engineered repositories. Since near-surface repositories are unsuitable for ILW, deep geological disposal is considered the most viable long-term solution for ILW. This method can effectively isolate the ILW from the biosphere and safeguards against environmental contamination.

2.3 High-level nuclear waste

HLW is the most hazardous category of radioactive waste, owing to its intense radioactivity and significant heat generation. This waste contains a mix of radioactive isotopes, including long-lived radionuclides that can remain hazardous for thousands to hundreds of thousands of years. HLW primarily consists of SNF removed from reactors after it has been used to generate electricity, as well as waste produced from reprocessing spent fuel. In the United States, the HLW also contains a large portion of defense waste, which was generated from the development of nuclear weapons and naval propulsion programs [7].

HLW primarily consists of fission products, transuranium elements, residual U, Pu, and activation products. These components make HLW highly radiotoxic and heat-generating, requiring long-term containment and isolation strategies to ensure environmental and public safety. For example, fission products such as ^{137}Cs, ^{90}Sr, and ^{129}I are highly radioactive and contribute significantly to short- to medium-term radiotoxicity of HLW. Transuranium elements, including ^{239}Pu, ^{241}Am, ^{237}Np, and Cm isotopes, are formed through neutron capture and subsequent beta decay [1, 8]. These isotopes dominate the long-term radiotoxicity due to their extended half-lives. Residual U and Pu left unreacted in the reactor, and the activation products like ^{60}Co and ^{63}Ni from neutron activation of reactor materials both contribute to the radioactivity of the HLW [1]. Considering the long half-lives of these isotopes, deep geological repositories (DGR) are considered the most viable solution.

3. Safe storage of nuclear waste

The management of nuclear waste requires an integrated approach that addresses both short- and long-term storage needs. The high radioactivity and varying half-lives of nuclear waste require adaptable and secure storage systems to protect human health and the environment.

3.1 Short-term storage solutions

SNF is extremely radioactive upon removal from the reactor, emitting lethal radiation levels and significant decay heat [1]. Although the thermal output declines to less than 1% within a century [9], SNF must be handled in heavily shielded environments. The SNF directly removed from the reactors are typically stored on-site at nuclear facilities where the waste is generated. This approach minimizes the immediate risks associated with transporting radioactive materials and allows facilities to maintain control over their waste. On-site storage often involves tailored facilities, such as cooling ponds (**Figure 3**) and dry cask storage (DCS, **Figure 4**) systems, which are

Figure 3.
Illustration of cooling pond. Source: U.S. NRC [10].

Figure 4.
Photograph of a dry cask storage site in the U.S. Source: U.S. NRC [10].

designed to manage the specific properties of the waste. Cooling ponds, also known as spent fuel pools, are the initial storage method for SNF [10]. These pools are constructed with reinforced concrete and lined with stainless steel (SS) to ensure structural integrity. The water also acts as a shield against radiation, protecting workers and the surrounding environment. On-site storage in cooling ponds can last several years, depending on the thermal and radiological characteristics of the waste.

As cooling ponds reach capacity, nuclear facilities transition spent fuel to DCS. This system involves encasing the spent fuel in thick-walled, airtight containers made of SS and concrete. These casks are stored above ground in secure, monitored areas designed to withstand external threats such as natural disasters or terrorist attacks. DCS provides a stable and passive cooling mechanism, eliminating the need for active systems and reducing operational costs. However, DCS was initially intended as a temporary measure before transferring nuclear fuel from storage pools to final disposal sites. It may now extend for up to a century, particularly in the United States, due to the absence of a geological repository to accommodate the spent fuel.

3.2 Long-term storage considerations

Long-term storage of nuclear waste presents significant challenges, as it must address the hazards associated with radioactive materials over timescales spanning hundreds of thousands of years. Geological repositories (**Figure 5**), also known as deep underground storage facilities, are widely regarded as the most viable solution for isolating high-level and some intermediate-level waste from the biosphere.

Geological repositories involve burying nuclear waste deep within stable geological formations, such as granite [12], salt domes [13, 14], or clay [15]. These formations provide natural barriers that prevent the transport of radionuclides into the environment. The design of such facilities incorporates multiple engineered barriers, including corrosion-resistant waste canisters, bentonite clay backfill, and reinforced tunnels. These barriers work in tandem with the surrounding geology to ensure containment over geological timescales.

Finland is the first country in the world to construct a DGR for SNF disposal. Their Onkalo facility is designed to house SNF for up to 100,000 years [16]. Other countries, such as Canada, France, and Sweden, are also actively developing

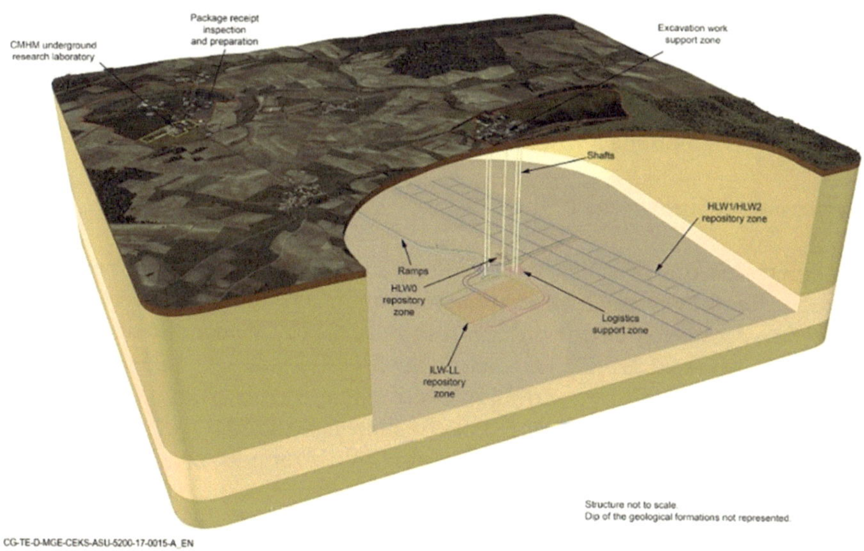

Figure 5.
Illustration of deep geological repository in France. Source: Andra (French national radioactive waste management agency) [11].

similar repositories. However, the implementation of geological repositories is often complicated by political and societal challenges. For example, the Yucca Mountain in the United States was initially proposed as the primary deep geological storage site, and it was codified into the federal law through the Nuclear Waste Policy Act of 1982. However, it has faced decades of political debates and public opposition, particularly from the state of Nevada due to the concerns about environmental safety and lack of democratic decision making [17–19]. The opponents of Yucca Mountain project also raised concerns about significant geological uncertainties, such as the faster water infiltration rates than expected and the discovery of colloidal transport of radionuclides [19]. These findings both undermine the suitability of the site as a dry repository. The shift in reliance from the natural geological features to engineered barriers led to further questions about the validity of the original scientific and policy decisions [19]. The project was ultimately suspended in 2010 due to political and local resistance, despite billions of dollars in investment and extensive scientific studies demonstrating the potential suitability of the site. Despite these setbacks, the lessons learned from Yucca Mountain and similar projects have informed new approaches to repository development. For example, Canada, Finland, and Sweden have prioritized community involvement and consent-based siting processes, which have proven more successful in gaining public acceptance.

3.3 Technical challenges of long-term containment

The primary technical challenge of long-term storage is ensuring the integrity of containment systems over geological timescales. Geological processes, such as earthquakes and water infiltration, can compromise repository stability. Additionally, the potential for human intrusion, whether intentional or accidental, remains a unique risk that must be addressed through proper site selection, design, monitor, and regulation. Materials used in containment systems must resist radiation, corrosion, and thermomechanical stresses, maintaining their integrity until the radionuclides decay to safe levels. Ensuring that these systems remain effective for tens to hundreds of thousands of years requires extensive research on experimental testing and computational simulations. Moreover, all proposed solutions should meet ethical requirements to avoid transferring undue burdens to future generations.

3.4 Improvements in storage casks

Advancements in storage cask technology have significantly enhanced the safety and efficiency of nuclear waste management. Modern dry casks are engineered to provide superior radiation shielding [20], thermal management [21], and structural durability [21]. After the SNF is loaded, the SS canisters are securely sealed either by welding or bolting to ensure airtight containment. Multilayer designs, incorporating materials such as borated SS [22], an effective neutron absorber, and high-density concrete [23], offer enhanced protection against radiation leakage and external threats. Advances in modular cask designs also allow for scalability and flexibility, enabling facilities to adapt to changing storage needs. Ongoing research focuses on addressing potential degradation mechanism [24], developing mitigation and repair techniques [25, 26], and creating advanced coatings to enhance durability [27].

3.5 Advances in containment materials

The development of advanced containment materials plays a critical role in improving the longevity and reliability of nuclear waste storage systems [3, 28]. Given the extreme complexity of the waste chemistry and the exceptionally long storage duration required on a geological time scale, a combination of material systems including metals, glass, and crystalline ceramics is employed [28–30].

Vitrification is the process of converting HLW into a stable, amorphous material by incorporating the complex waste into a molten glass matrix, which solidifies into a durable form suitable for long-term storage and disposal. It is a widely adopted method for stabilizing HLW for many countries including the France, UK, and US [31–33]. In this process, radioactive materials are mixed with molten glass and poured into metallic canisters, where the glass solidifies [34]. The resulting glass matrix effectively immobilizes radionuclides, preventing their release into the environment.

Vitrification offers several advantages. For example, the glass matrix is highly durable and resistant to water and radiation damage, ensuring long-term containment [28]. The process also significantly reduces the volume of waste, making storage and transportation more manageable. Vitrified waste can be safely stored in interim facilities or incorporated into geological repositories for permanent disposal. Despite its benefits, vitrification is a highly intricate process, largely due to the complex chemical composition of HLW, which poses significant challenges in designing stable and durable glass formulations [35]. Additionally, operational difficulties such as controlling spinel crystallization, managing noble metal accumulation, and addressing refractory corrosion in melters significantly impact the efficiency and longevity of vitrification systems [35].

Metallic canisters for the long-term storage of HLW are made from various materials depending on the country. In the US, SS is commonly used, while carbon steel is preferred in the Belgian and French programs [36]. Countries like Canada, Finland, and Sweden are developing a copper-coated steel canister [37]. In addition, high-performance alloys, such as Hastelloy, Inconel alloys, and titanium alloys, which are resistant to corrosion, radiation damage, and thermal stresses are being considered as engineered barrier materials [38].

In addition to glass, a broad range of crystalline ceramic materials were also proposed as a superior medium for immobilizing HLW by incorporating radioactive elements into the crystal lattices of stable mineral phases [39, 40]. Compared to borosilicate glass, several ceramics demonstrate enhanced durability, resistance to leaching, and long-term stability in geological environments, offering a cost-effective and reliable solution for nuclear waste containment [39].

Bentonite clay, a natural material with excellent sealing and self-healing properties, is widely used as a backfill in geological repositories to provide an additional barrier against the transport of radionuclide [41]. Innovative coatings [42] and composites [43] are also being explored to enhance the durability of storage canisters and repository structures. These materials are designed to maintain their integrity under extreme conditions, ensuring the long-term containment of radioactive waste.

3.6 Monitoring and predictive technologies

The development of advanced monitoring technologies has revolutionized the management of nuclear waste storage facilities. For example, remote monitoring systems integrating nondestructive analysis and containment/surveillance techniques were developed to continuously monitor and transmit data in the DCS [44, 45]. These systems enable verification of SNF assemblies inside sealed dry casks without opening them, ensuring security and reducing inspection needs. In addition, efforts have also been directed toward advancing digitalization in nuclear waste management through the adoption of cutting-edge technologies such as digital twins, machine learning, and high-performance computing, aiming to enhance safety, optimize efficiency, and refine predictive modeling for waste storage and disposal [46]. These tools integrate data from geological, geotechnical, and monitoring systems to enhance decision-making, optimize repository designs, and ensure long-term environmental protection.

Predictive models, including physics-based and data-driven, are also being employed to simulate the long-term behavior of storage systems under various scenarios [47–56]. By analyzing factors such as geological shifts, material degradation, and climate change, these models help refine storage designs and mitigate risks. Artificial intelligence and machine learning are further enhancing predictive capabilities, enabling more accurate and efficient decision-making [28, 57, 58].

4. Other long-term disposal methods

While DGR is widely regarded as the most reliable method for long-term nuclear waste disposal, several other methods were also explored by different countries. Some of these alternatives are illustrated in this section including deep borehole disposal, transmutation technologies, and other methods, including oceanic and space disposal.

4.1 Deep borehole disposal

The deep borehole disposal (DBD) method involves drilling boreholes to depths of 3–5 kilometers in stable crystalline rock formations [59]. These depths ensure isolation from surface environments and groundwater systems for millions of years, leveraging the natural geological barriers of the Earth. In this process, waste canisters are loaded into the lower sections of the borehole, typically the bottom 1–2 kilometers. These waste-filled sections are then sealed with multiple engineered barriers, including bentonite clay, cement, and other materials designed to prevent waste migration. The upper portions of the borehole are filled with impermeable seals (e.g., graphite, silica sand, bentonite, or molten lead) to further secure the containment.

Compared to DGR, DBD minimizes land usage while reducing risks of human intrusion. It also avoids the need for extensive surface facilities and therefore substantially reduces carbon footprint. Additionally, it targets geologically stable areas remote from population centers, thereby ensuring enhanced safety. DBD is also considered a cost-effective and secure solution for countries with limited waste inventories or those seeking decentralized disposal options. However, DBD also comes with its own challenges. For example, it is difficult to maintain the stability of large diameter boreholes at depths of 4–5 kilometers, where geomechanical stresses can cause elongation or collapse [60]. Additionally, the design and deployment of effective sealing and support systems for waste packages in such extreme conditions require advanced materials and techniques to ensure long-term containment and safety.

4.2 Transmutation

Transmutation is a process converting long-lived radioactive isotopes, such as ^{239}Pu and ^{241}Am, into shorter-lived or stable isotopes through nuclear reactions [61]. It was proposed as an effective technology to reduce the radioactivity and longevity of nuclear waste by transforming HLW into less hazardous forms. The process typically occurs in specialized reactors or accelerator-driven systems. Fast neutron reactors, for instance, can induce fission in long-lived isotopes, breaking them down into shorter-lived fission products. Similarly, particle accelerators can bombard isotopes with protons or neutrons to achieve the same effect.

Transmutation technologies are still in the experimental stage and face several challenges. The major challenges include the technological difficulties of managing the heat and neutron source from transuranics during fuel recycling and addressing the substantial costs and safety concerns associated with fuel handling and fabrication [62]. Additionally, despite its potential to reduce radiotoxicity and waste volume, recent studies indicate that the influence of transmutation on geological disposal strategies is relatively modest [62]. Achieving widespread adoption will require significant advancements in reactor and fuel technologies.

4.3 Other disposal methods

In the mid-twentieth century, oceanic disposal was considered a viable option for managing nuclear waste [63]. Radioactive materials were encased in drums and dumped into deep ocean trenches [64], where proponents argued they would remain isolated due to the immense pressure and stable conditions. However, concerns about leakage, environmental contamination, and the ethical implications of polluting global ecosystems led to the practice being largely banned by international agreements, such as the London Convention of 1970s [65]. Oceanic disposal is no longer pursued as a legitimate waste management strategy, but it remains a cautionary tale about the importance of thoroughly evaluating environmental impacts.

Space disposal involves launching nuclear waste into outer space, typically by sending it toward the Sun or beyond Earth's orbit [66]. This method in theory offers complete isolation of radioactive materials from the planet. However, the costs and risks associated with space disposal make it impractical. Launch failures, which are not uncommon in the current era, could spread radioactive materials across the atmosphere of the Earth, resulting in catastrophic contamination. The energy requirements and logistical challenges of launching heavy waste payloads also substantially limit the feasibility of this method.

5. Environmental and health risks

The management of nuclear waste is important to safeguard the environment and public health. Improper disposal of radioactive materials can lead to significant risks, including environmental contamination and severe health impacts.

5.1 Potential risks associated with improper disposal

The failure of storage systems and the subsequent release of radionuclides into the environment can lead to profound, long-lasting consequences.

The significance of different radionuclides is influenced by their mobility under the geochemical conditions of the disposal site. Elements like iodine are highly mobile and readily transport through the natural barrier, whereas actinides remain largely immobile under reducing conditions typical of DGR [67]. One of the most concerning risks is groundwater contamination. Radioactive isotopes, if not securely contained, can leach into groundwater supplies, spreading irreversible contamination over vast areas. Once radionuclides enter aquifers, they are exceedingly difficult to remove and can remain hazardous for thousands of years. Contaminated groundwater not only affects ecosystems but also poses significant risks to communities reliant on these water sources for drinking, agriculture, and industry. Given the critical importance, the degradation mechanism of SNF and vitrified HLW in contact with groundwater has been the focus of numerous ongoing studies [28, 33, 52, 68–70].

Another major risk is the release of radioactive materials into the atmosphere or surface environment. Accidental breaches in storage systems, such as those caused by natural disasters, human error, or sabotage, could result in radioactive particles being dispersed into the air or onto the ground. Such events can lead to widespread environmental contamination, rendering areas uninhabitable and causing severe ecological damage. Wildlife and plant life exposed to radiation often suffer mutations, reproductive failures, and population declines, further destabilizing ecosystems.

5.2 Health impacts

Exposure to radioactive isotopes can lead to significant health risks to humans, primarily categorized into somatic and genetic effects [71, 72]. Somatic effects occur immediately or shortly after high-dose radiation exposure, causing cellular damage, tissue destruction, and organ failure. Acute exposure to high doses of radiation ($\geq$1 Gy) in a short period can cause acute radiation sickness, characterized by nausea, vomiting, hair loss, and damage to internal organs [73]. In extreme cases, it can lead to death within days or weeks [73].

Genetic effects manifest over time and result from DNA damage, increasing the risk of mutations and carcinogenesis, potentially affecting future generations. Long-term exposure, even at lower levels, carries significant risks, particularly in the form of cancer [74]. Radioactive isotopes such as ^{137}Cs and ^{90}Sr can be inhaled or ingested, becoming incorporated into the body and irradiating tissues over time. This increases the likelihood of developing cancers, including leukemia, thyroid cancer, and bone cancer [72]. These risks vary depending on the type of exposure, dose, and individual factors, such as age and pre-existing conditions like iodine deficiency. Pregnant women and children are especially susceptible to the effects of radiation, as their cells are dividing more rapidly. For example, children exposed to ^{131}I are especially vulnerable to thyroid cancer [72]. Other health effects, such as cardiovascular disease and potential genetic damage, may also manifest over time [72].

6. Legal and regulatory framework

The safe management of nuclear waste is a global concern that requires cohesive international guidelines, robust national regulations, and stringent licensing and monitoring protocols.

6.1 International guidelines and agreements

The International Atomic Energy Agency (IAEA) plays a central role in establishing global standards for nuclear waste management. As the world's "Atoms for Peace and Development" organization, the IAEA provides technical guidelines for the safe handling of radioactive materials. It facilitates capacity building through training and peer reviews while promoting international collaboration and research in waste management technologies. Additionally, the IAEA conducts assessments and offers recommendations to improve waste management practices. The Non-Proliferation Treaty (NPT) also contributes to nuclear waste management by promoting the peaceful use of nuclear technology. Signatory countries under the NPT commit to transparency in their nuclear programs, ensuring that radioactive materials are used safely and securely. Various regional agreements and conventions, such as the Joint Convention on the Safety of Spent Fuel Management and on the Safety of Radioactive Waste Management, also facilitate international cooperation and shared accountability.

6.2 National regulations

Different countries have tailored approaches to nuclear waste management, reflecting their unique energy policies and geographic conditions. Each country has its own agency responsible for developing regulations and conducting inspections/licensing to ensure that operators meet stringent safety criteria. Their oversight is critical to maintaining public confidence in nuclear energy. For example, the Nuclear Regulatory Commission (NRC) oversees the storage and disposal of nuclear waste in the United States. The NRC enforces strict licensing processes, requiring facilities to demonstrate compliance with safety standards (e.g., Code of Federal Regulations). Although the Yucca Mountain project was suspended, DCS and interim storage sites are regulated rigorously to ensure environmental protection and public safety. As a global leader in nuclear energy, France operates an integrated system for nuclear waste management, including reprocessing spent fuel to recover usable materials and centralized storage of intermediate and high-level waste. The National Radioactive Waste Management Agency (ANDRA) is developing a DGR in Bure known as Cigéo, which exemplifies France's commitment to long-term disposal solutions. Japan's approach to nuclear waste management has been impacted by the Fukushima disaster, leading to heightened regulatory oversight. The Nuclear Regulation Authority enforces strict safety standards, particularly for interim storage and the future development of geological repositories. Public trust remains a critical challenge, requiring ongoing community engagement and transparency [75].

6.3 Licensing and monitoring requirements

Licensing is a critical part of nuclear waste management to ensure facilities meet safety and environmental standards before operations begin. Licensing in most countries involves a comprehensive review process, including environmental impact assessments, safety analyses, and public consultations. Once a facility is operational, periodic monitoring is conducted to ensure continued compliance. Regulatory bodies, such as the NRC in the United States and the IAEA internationally, mandate regular inspections and performance reviews. Operators are required to maintain detailed records of waste inventories and storage conditions. These records are subject to review by regulatory authorities to verify compliance and ensure transparency. The IAEA also facilitates international peer reviews through programs such as the Integrated Review Service for Radioactive Waste and Spent Fuel Management, Decommissioning, and Remediation (ARTEMIS) and provides additional oversight, allowing countries to learn from each other and adopt best practices.

7. Future directions

As technology advances and global priorities shift toward environmental sustainability, the future of nuclear waste management will strongly depend on integrating cutting-edge solutions, embracing sustainability principles, and fostering transparent engagement with the public.

7.1 Innovations in nuclear waste management

The development of new containment materials and strategies continues to enhance the safety and reliability of nuclear waste storage. Advanced materials such as high-performance glass, crystalline ceramics, alloys, composites, and robust coatings are being designed to resist corrosion, radiation damage, and thermal stresses over millennia. Future technologies, such as the Generation IV reactors, also promise to reshape the nuclear waste landscape. For example, the molten salt reactors (MSRs) offer significant benefits by efficiently utilizing fuel, including Th and recycled nuclear waste, thereby reducing the volume and radiotoxicity of nuclear waste [76]. Additionally, their ability to operate at higher thermal efficiencies and support online refueling enhances sustainability and minimizes long-lived waste compared to traditional reactor technologies [77].

7.2 Sustainability and circular economy in nuclear waste

A key direction in nuclear waste management is aligning with the principles of sustainability and the circular economy. Reprocessing nuclear waste to extract valuable isotopes like Pu and U, combined with vitrification, can significantly reduce the volume of HLW while providing materials for reuse in reactors [78]. France, for example, has successfully implemented reprocessing at scale, recovering fissile materials for use in mixed-oxide fuel. Other countries, including China, Germany, Japan, and UK, also reprocess their waste [79].

7.3 Public engagement and transparency

Public acceptance remains a critical factor in the success of nuclear waste management initiatives. Educating the public about the science, risks, and benefits of nuclear energy and waste disposal is essential for fostering trust and addressing concerns. Transparent communication about safety measures, site selection processes, and long-term plans can alleviate fears and build confidence in nuclear projects. Engaging local communities in decision-making processes is particularly important. Successful projects, such as the Onkalo repository in Finland, demonstrate the value of involving stakeholders early and maintaining open dialog [79]. Additionally, investing in youth education and public campaigns to demystify nuclear technology ensures that future generations are equipped to make informed decisions.

8. Conclusion

Nuclear waste management is one of the most complex challenges associated with the use of nuclear energy. Throughout this chapter, we have explored various aspects of this issue. The primary concern is linked to the safe isolation of hazardous materials from human activity over timescales that span tens to hundreds of thousands of years. During such a long period, the safety of human beings and the global environment can be protected through an integrated approach combining short-term storage solutions like cooling ponds and dry casks, and long-term strategies such as DGRs. Technological advances in containment materials, reactor designs, and monitoring sensors are also continuously improving the safety and efficiency of waste management. These strategies are complemented by effective regulatory frameworks and public engagement. By embracing technical innovation and fostering trust, the global community can address the challenges of nuclear waste while ensuring a cleaner, safer energy future for generations to come.

Acknowledgements

The time of XG spent drafting this book chapter was supported by the Nuclear Energy University Program, US Department of Energy, through Award No. NE0009388.

Conflict of interest

The author declares no conflict of interest.

Author details

Xiaolei Guo
Department of Metallurgical and Materials Engineering, Colorado School of Mines, Golden, CO, USA

*Address all correspondence to: xiaolei.guo@mines.edu

References

[1] Ewing RC. Long-term storage of spent nuclear fuel. Nature Materials. 2015;**14**(3):252-257

[2] Felveson H. Managing Spent Fuel from Nuclear Power Reactors: Experience and Lessons from around the World. Princeton, NJ, USA: International Panel on Fissile Materials; 2011

[3] Ojovan MI, Lee WE, Kalmykov SN. An Introduction to Nuclear Waste Immobilisation. Amsterdam, Netherlands: Elsevier; 2019

[4] Cui D, Low J, Sjöstedt CJ, Spahiu K. On Mo-Ru-Tc-Pd-Rh-Te alloy particles extracted from spent fuel and their leaching behavior under Ar and H2 atmospheres. Radiochimica Acta. 2004;**92**(9-11):551-555

[5] O'boyle D, Brown F, Dwtght A. Analysis of fission product ingots formed in uranium-plutonium oxide irradiated in EBR-II. Journal of Nuclear Materials. 1970;**35**(3):257-266

[6] Strachan D et al. Epsilon metal: A waste form for noble metals from used nuclear fuel. Journal of Nuclear Materials. 2020;**532**:152040

[7] Ewing RC, Weber WJ, Clinard FW Jr. Radiation effects in nuclear waste forms for high-level radioactive waste. Progress in Nuclear Energy. 1995;**29**(2):63-127

[8] Baybarz R. Recovery and application of the transuranium elements 237 Np, 238 Pu, 241 Am, 242 Cm, 244 Cm and 252 Cf. Atomic Energy Review. 1970;**8**(2):327-360

[9] Ansolabehere SD, Driscoll M, Holdren JP, Joskow PL, Lester RK, Moniz EJ, et al. The Future of Nuclear Power: An Interdisciplinary MIT Study. Cambridge, MA, USA: Massachusetts Institute of Technology; 2003

[10] U.S. NRC. Information Digest. NUREG-1350, Section 4. Vol. 342022. p. 6

[11] Delay J. Synthesis of 20 Years Research, Development and Demonstration in Andra's Underground Research Laboratory in Bure for Cigéo Project-France. Andra Technical Document. ID: DRPAS3. 2019

[12] Witherspoon P et al. Rock mass characterization for storage of nuclear waste in granite. In: ISRM Congress. Montreux, Switzerland: ISRM; 1979

[13] von Berlepsch T. Salt repository systems: Design development approach at the example of the Gorleben salt dome. In: Geological Repository Systems for Safe Disposal of Spent Nuclear Fuels and Radioactive Waste. Duxford, UK: Elsevier; 2017. pp. 145-162

[14] Martinez J, Thoms R, Kupfer D. Investigation of the Utility of Gulf Coast Salt Domes for the Storage or Disposal of Radioactive Wastes. Baton Rouge, LA, USA: Louisiana State University. Institute for Environmental Studies; 1976

[15] Hansen F, Hardin E, Orrell A. Clay Repository Research in the USA. International Nuclear Information System, Paper Number: INIS-FR-15-Clays-Nantes-2010, Nantes, France; 2010

[16] Siren T. Overview of finnish spent nuclear fuel disposal programme. Journal of the Korean Society of Mineral and Energy Resources Engineers. 2017;**54**(4):367-376

[17] Winograd IJ, Roseboom EH Jr. Yucca Mountain Revisited. Washington, DC, USA: American Association for the Advancement of Science; 2008. pp. 1426-1427

[18] Flynn J et al. Trust as a determinant of opposition to a high-level radioactive waste repository: Analysis of a structural model. Risk Analysis. 1992;**12**(3):417-429

[19] Macfarlane A. Underlying Yucca Mountain: The interplay of geology and policy in nuclear waste disposal. Social Studies of Science. 2003;**33**(5):783-807

[20] Gao Y et al. Dry cask radiation shielding validation and estimation of cask surface dose rate with MAVRIC during long-term storage. Annals of Nuclear Energy. 2020;**140**:107145

[21] Li J, Liu YY. Thermal modeling of a vertical dry storage cask for used nuclear fuel. Nuclear Engineering and Design. 2016;**301**:74-88

[22] He X, Ahn T, Sippel T. Corrosion of borated stainless steel in water and humid air. In: NACE Corrosion. Houston, TX, USA: NACE; 2012

[23] Lessing P. High-density and radiation-shielding concrete. In: Developments in the Formulation and Reinforcement of Concrete. Cambridge, UK: Woodhead Publishing; 2008. pp. 44-78

[24] Bryan CR, Enos D. FY16 Status Report: SNF Interim Storage Canister Corrosion and Surface Environment Investigations. Albuquerque, NM, United States: Sandia National Lab. (SNL-NM); 2017

[25] Qu HJ, Tatman J, Wharry JP. Chloride-induced stress corrosion cracking in austenitic steels for SNF storage canisters-recent understanding and advances in mitigation and repair. Journal of Nuclear Materials. 2024;**596**:155080

[26] Ross KA et al. Investigation of Cold Spray as a Dry Storage Canister Repair and Mitigation Tool. Richland, WA, United States: Pacific Northwest National Lab. (PNNL); 2020

[27] Knight A et al. FY21 Status Report: SNF Canister Coatings for Corrosion Prevention and Mitigation. Sandia National Laboratory, Albuquerque, NM, USA: US Department of Energy; 2021

[28] Frankel GS et al. Recent advances in corrosion science applicable to disposal of high-level nuclear waste. Chemical Reviews. 2021;**121**(20):12327

[29] Frankel GS et al. A comparative review of the aqueous corrosion of glasses, crystalline ceramics, and metals. npj Materials Degradation. 2018;**2**(1):15

[30] Guo X et al. Reply to: How much does corrosion of nuclear waste matrices matter. Nature Materials. 2020;**19**:962-963

[31] Vienna JD. Nuclear waste vitrification in the United States: Recent developments and future options. International Journal of Applied Glass Science. 2010;**1**(3):309-321

[32] Gin S et al. Radionuclides containment in nuclear glasses: An overview. Radiochimica Acta. 2017;**105**(11):927-959

[33] Gin S et al. An international initiative on long-term behavior of high-level nuclear waste glass. Materials Today. 2013;**16**(6):243-248

[34] Guo X et al. Self-accelerated corrosion of nuclear waste forms at material interfaces. Nature Materials. 2020;**19**:310-316

[35] Goel A et al. Challenges with vitrification of Hanford high-level waste (HLW) to borosilicate glass—An overview. Journal of Non-Crystalline Solids X. 2019;**4**:100033

[36] Guo X, Gin S, Frankel GS. Review of corrosion interactions between different materials relevant to disposal of high-level nuclear waste. npj Materials Degradation. 2020;**4**(1):1-16

[37] King F, Lilja C, Vähänen M. Progress in the understanding of the long-term corrosion behaviour of copper canisters. Journal of Nuclear Materials. 2013;**438**(1-3):228-237

[38] Rebak R, Shen T, Summers T. Prediction of the Lifetime Integrity of a Nuclear Waste Container Material Based on Thermal Stability Studies. Livermore, CA, United States: Lawrence Livermore National Lab. (LLNL); 1999

[39] Ringwood AE et al. Immobilisation of high level nuclear reactor wastes in SYNROC. Nature. 1979;**278**(5701):219-223

[40] Ewing RC, Lutze W. High-level nuclear waste immobilization with ceramics. Ceramics International. 1991;**17**(5):287-293

[41] Gates WP, Bouazza A, Churchman GJ. Bentonite clay keeps pollutants at bay. Elements. 2009;**5**(2):105-110

[42] Choi H-J, Lee M, Lee JY. Application of a cold spray technique to the fabrication of a copper canister for the geological disposal of CANDU spent fuels. Nuclear Engineering and Design. 2010;**240**(10):2714-2720

[43] Evarts JS et al. Ceramic–metal (cermet) composites: A review of key properties and synthesis methods focused on nuclear waste immobilization. Industrial & Engineering Chemistry Research. 2024;**63**(14):6003-6023

[44] Sagadevan A, Chirayath S. Information driven safeguards approach for remote monitoring system of dry cask storage. Nuclear Instruments and Methods in Physics Research Section A: Accelerators, Spectrometers, Detectors and Associated Equipment. 2020;**954**:161737

[45] Jonkmans G et al. Nuclear waste imaging and spent fuel verification by muon tomography. Annals of Nuclear Energy. 2013;**53**:267-273

[46] Kolditz O et al. Digitalisation for nuclear waste management: Predisposal and disposal. Environmental Earth Sciences. 2023;**82**(1):42

[47] Lutze W et al. Alteration of basalt glasses: Implications for modelling the long-term stability of nuclear waste glasses. Nature. 1985;**314**(6008):252-255

[48] Macdonald D, Urquidi-Macdonald M. Thin-layer mixed-potential model for the corrosion of high-level nuclear waste canisters. Corrosion. 1990;**46**(5):380-390

[49] Ewing RC et al. Performance assessments of nuclear waste repositories: A dialogue on their value and limitations. Risk Analysis. 1999;**19**(5):933-958

[50] Rechard RP. Historical relationship between performance assessment for radioactive waste disposal and other types of risk assessment. Risk Analysis. 1999;**19**(5):763-807

[51] Dunn D, Cragnolino G, Sridhar N. An electrochemical approach to predicting long-term localized corrosion of corrosion-resistant high-level

waste container materials. Corrosion. 2000;**56**(01):90-104

[52] Vienna JD et al. Current understanding and remaining challenges in modeling long-term degradation of borosilicate nuclear waste glasses. International Journal of Applied Glass Science. 2013;**4**(4):283-294

[53] Rechard RP, Wilson ML, Sevougian SD. Progression of performance assessment modeling for the Yucca Mountain disposal system for spent nuclear fuel and high-level radioactive waste. Reliability Engineering & System Safety. 2014;**122**:96-123

[54] Hansen CW et al. Overview of total system model used for the 2008 performance assessment for the proposed high-level radioactive waste repository at Yucca Mountain, Nevada. Reliability Engineering & System Safety. 2014;**122**:249-266

[55] Herranz LE, Penalva J, Feria F. CFD analysis of a cask for spent fuel dry storage: Model fundamentals and sensitivity studies. Annals of Nuclear Energy. 2015;**76**:54-62

[56] Lee C-J, Lee KJ. Application of Bayesian network to the probabilistic risk assessment of nuclear waste disposal. Reliability Engineering & System Safety. 2006;**91**(5):515-532

[57] Ben-Haim M, Macdonald DD. Modeling geological brines in salt-dome high level nuclear waste isolation repositories by artificial neural networks. Corrosion Science. 1994;**36**(2):385-393

[58] Hu G, Pfingsten W. Data-driven machine learning for disposal of high-level nuclear waste: A review. Annals of Nuclear Energy. 2023;**180**:109452

[59] Brady PV et al. Deep Borehole Disposal of High-Level Radioactive Waste. Sandia Report SAND2009-4401. Albuquerque, NM: Sandia National Laboratories; 2009

[60] Beswick AJ, Gibb FG, Travis KP. Deep borehole disposal of nuclear waste: Engineering challenges. Proceedings of the Institution of Civil Engineers: Energy. 2014;**167**(2):47-66

[61] National Research Council, Nuclear Wastes: Technologies for Separations and Transmutation. Washington, DC, USA: National Academy Press; 1996

[62] Salvatores M, Palmiotti G. Radioactive waste partitioning and transmutation within advanced fuel cycles: Achievements and challenges. Progress in Particle and Nuclear Physics. 2011;**66**(1):144-166

[63] Pritchard DW. Disposal of radioactive wastes in the ocean. Health Physics. 1961;**6**(2):103-109

[64] Calmet DP. Ocean disposal of radioactive waste. Status report. IAEA Bulletin. 1989;**31**(4):47-50

[65] de La Fayette L. The London convention 1972: Preparing for the future. The International Journal of Marine and Coastal Law. 1998;**13**(4):515-536

[66] Burns R et al. Nuclear Waste Disposal in Space. Washington, DC, USA: NASA; 1978

[67] Grambow B. Mobile fission and activation products in nuclear waste disposal. Journal of Contaminant Hydrology. 2008;**102**(3-4):180-186

[68] Shoesmith DW. Fuel corrosion processes under waste disposal conditions. Journal of Nuclear Materials. 2000;**282**(1):1-31

[69] Hanson BD et al. Corrosion of commercial spent nuclear fuel. 1. Formation of studtite and metastudtite. Radiochimica Acta. 2005;**93**(3):159-168

[70] Forsyth RS, Werme LO. Spent fuel corrosion and dissolution. Journal of Nuclear Materials. 1992;**190**:3-19

[71] Yamamoto LG. Risks and management of radiation exposure. Pediatric Emergency Care. 2013;**29**(9):1016-1026

[72] Christodouleas JP et al. Short-term and long-term health risks of nuclear-power-plant accidents. New England Journal of Medicine. 2011;**364**(24):2334-2341

[73] Donnelly EH et al. Acute radiation syndrome: Assessment and management. Southern Medical Journal. 2010;**103**(6):541

[74] Kamiya K et al. Long-term effects of radiation exposure on health. The Lancet. 2015;**386**(9992):469-478

[75] Miller DS. Public trust in the aftermath of natural and na-technological disasters: Hurricane Katrina and the Fukushima Daiichi nuclear incident. International Journal of Sociology and Social Policy. 2016;**36**(5/6):410-431

[76] Rykhlevskii A et al. Fuel Cycle Performance of Fast Spectrum Molten Salt Reactor Designs. Oak Ridge, TN, United States: Oak Ridge National Laboratory (ORNL); 2019

[77] Serp J et al. The molten salt reactor (MSR) in generation IV: Overview and perspectives. Progress in Nuclear Energy. 2014;**77**:308-319

[78] Caruso S, Meleshyn A, Noseck U. Estimation and comparison of the radionuclide inventories in vitrified high-level wastes from reprocessing plant. Progress in Nuclear Energy. 2017;**94**:216-221

[79] Kurniawan TA et al. Technological solutions for long-term storage of partially used nuclear waste: A critical review. Annals of Nuclear Energy. 2022;**166**:108736

Section 2

Fission Applications

Chapter 4

Merchant Nuclear-Powered Vessels: A New Frontier in Maritime Propulsion

Nima Fathi and Milton Korn

Abstract

The maritime industry is at a pivotal moment, grappling with the need to reduce greenhouse gas emissions while maintaining operational efficiency. Traditional fossil-fuel-based propulsion systems are major contributors to environmental pollution. In response, the industry is investigating alternative energy sources to meet stringent international regulations and adopt sustainable practices. Among these, nuclear propulsion emerges as a compelling solution, offering unmatched energy density, extended operational range, and near-zero operational emissions. This chapter provides a comprehensive exploration of the technical, regulatory, and economic dimensions of deploying nuclear technology in merchant vessels, highlighting its transformative potential for maritime propulsion. Focusing on advancements in ship propulsion technologies, this chapter examines modern nuclear reactor designs, fuel cycles, and operational economics. Historical case studies, such as the NS Savannah, are revisited to draw insights for contemporary applications. Key topics include maneuverability, cost analysis, refueling strategies, and waste management, with a detailed review of advanced reactor technologies like Generation IV Molten Salt Reactors (MSRs) and Lead-cooled Fast Reactors (LFRs). Additionally, the chapter evaluates the potential of turbo-electric propulsion systems for commercial shipping. Through this analysis, it situates nuclear propulsion as a viable and sustainable alternative for the future of maritime operations.

Keywords: merchant vessels, maritime decarbonization, advanced reactor technologies, regulatory frameworks, sustainable shipping

1. Introduction

The application of nuclear propulsion in maritime transportation has been a subject of ongoing interest and debate, offering the potential to revolutionize the industry through enhanced efficiency, reduced environmental impact, and unparalleled operational range. While the concept is not new, with the first operational example being the NS Savannah, technological advancements and evolving global priorities have rekindled interest in nuclear-powered ships. The Savannah's operational history offers invaluable insights into the benefits and challenges of nuclear propulsion, but it must be interpreted within the context of mid-twentieth-century technology, economics, and regulatory frameworks. As the maritime industry seeks sustainable alternatives to conventional propulsion, the integration of nuclear power emerges as a compelling solution.

This chapter explores the historical evolution and modern resurgence of nuclear propulsion technologies, with a focus on bridging past lessons with future possibilities. The discussion begins by revisiting the NS Savannah's legacy, not merely as the first nuclear-powered merchant ship but as a critical milestone that laid the groundwork for advanced maritime reactor designs. While the Savannah demonstrated the feasibility of nuclear propulsion, its challenges—particularly in cost, refueling logistics, and waste management—highlight areas for improvement in contemporary designs.

Technological advancements since the Savannah era have expanded the potential for nuclear propulsion. Reactor designs such as pressurized water reactors (PWRs), lead-cooled fast reactors (LFRs), and molten salt reactors (MSRs) have matured significantly, offering diverse capabilities suited to varying maritime applications. For example, LFRs and MSRs, categorized as Generation IV reactors, bring extended operational lifespans, enhanced fuel utilization, and reduced refueling requirements, making them particularly attractive for long-haul and remote operations. The adaptability of these advanced designs allows for innovative approaches to addressing key maritime challenges such as maneuverability, cost efficiency, safety, and regulatory compliance [1–4].

One of the most critical considerations in nuclear propulsion is its ability to adapt to the dynamic operational requirements of maritime vessels. Unlike terrestrial reactors, which operate continuously at steady output, ship propulsion systems must respond to highly variable power demands, especially during maneuvers, docking, or emergency operations. Turbo-electric propulsion systems, which decouple the reactor's power output from immediate propulsion needs, offer a promising solution to these challenges by providing flexibility and improving fuel efficiency. This chapter examines how such systems, along with modern reactor technologies, can address these unique requirements [5].

Equally important is the economic viability of nuclear propulsion. While the high initial capital costs of nuclear reactors have historically deterred widespread adoption in the commercial sector, advancements in modular reactor design, fuel efficiency, and operational lifespan are reshaping cost dynamics. Small modular reactors (SMRs) and reactors with extended refueling intervals have the potential to reduce operational expenses and align nuclear propulsion with the financial realities of the shipping industry [6, 7].

Environmental considerations are also central to this discussion. Nuclear propulsion offers the dual advantage of near-zero greenhouse gas emissions during operation and reduced dependence on conventional fossil fuels. However, the management of nuclear waste remains a critical challenge. Advanced reactors, such as MSRs, offer innovative solutions like continuous reconditioning of liquid fuel to minimize waste generation, while long-life reactor cores, such as those in LFRs, can synchronize reactor lifespans with vessel lifespans, eliminating mid-life refueling needs [1–4, 8].

This chapter also considers the broader implications of integrating nuclear propulsion into the maritime sector, including regulatory, safety, and public perception challenges. Drawing on historical experiences, contemporary technological developments, and future-oriented concepts, it aims to provide a comprehensive framework for evaluating the feasibility and sustainability of nuclear-powered shipping [9–11].

2. Maneuverability of nuclear-powered vessels: Managing power near shore and offshore

The ability of nuclear-powered vessels to adapt their power output efficiently to varying operational demands is one of the most critical challenges in maritime nuclear propulsion. Unlike conventional marine propulsion systems powered by diesel engines or gas turbines, nuclear reactors are optimized for steady-state operation. They excel at producing constant power for extended periods, making them ideal for open-sea navigation. However, maritime operations, especially those near shore or during docking, require frequent power adjustments, including rapid deceleration, acceleration, and fine control for safe navigation in constrained spaces. Achieving this level of maneuverability while maintaining reactor safety and operational efficiency is vital for the practical implementation of nuclear propulsion in commercial and military shipping.

Managing power variability in nuclear-powered vessels involves two main approaches: adjusting the reactor's intrinsic operating parameters and integrating external systems to buffer power demands. Traditional nuclear reactors regulate their power output by adjusting the position of control rods within the reactor core. These control rods absorb neutrons, thereby moderating the fission process and controlling the reactor's heat generation. By varying the position of the control rods, reactor operators can align the reactor's output with propulsion needs. However, this approach has limitations. Frequent and rapid adjustments impose significant thermal and mechanical stresses on the reactor components, potentially reducing their lifespan and compromising long-term reliability. Modern advancements in reactor design aim to mitigate these effects by incorporating passive safety features and materials with greater thermal tolerance [11].

In addition to intrinsic adjustments, integrating energy storage systems has emerged as a promising strategy to manage the variable power demands of nuclear-powered ships. These systems act as intermediaries, decoupling the reactor's steady-state operation from the dynamic requirements of the propulsion system. During periods of low power demand, excess energy produced by the reactor can be stored and later released to meet short-term power spikes, such as those encountered during docking or sharp maneuvers. Various forms of energy storage, including batteries, flywheels, and thermal storage systems, have been explored for this purpose. Batteries store electrical energy, providing rapid and precise power delivery, while flywheels store kinetic energy and offer quick-response capabilities. Thermal storage systems, on the other hand, capture excess heat generated by the reactor, which can be converted to propulsion energy when needed. These technologies not only enhance maneuverability but also alleviate stress on the reactor, ensuring consistent performance over time.

Historical precedents provide valuable insights into how nuclear-powered vessels have addressed maneuverability challenges. The NS Savannah, the world's first nuclear-powered merchant ship, relied on control rod adjustments to meet varying power demands during docking and near-shore operations. While this method demonstrated the feasibility of nuclear propulsion for maritime use, it also revealed its limitations, particularly in the ship's ability to handle rapid fluctuations in power demand. The technology of the era constrained the Savannah's capacity to provide the precision needed for complex maneuvers [12, 13].

Military submarines, which have employed nuclear propulsion for decades, offer more advanced examples of maneuverability. These vessels frequently operate in challenging environments, such as shallow waters and intricate underwater terrains, where precise power control is essential. Submarines achieve this by combining advanced reactor designs with sophisticated control systems that allow real-time adjustments to reactor output while maintaining overall operational efficiency. Russian Alfa-class submarines, powered by lead-bismuth eutectic (LBE) reactors, exemplify the advantages of innovative reactor designs. Their reactors' high thermal efficiency and rapid heat dissipation capabilities allowed these submarines to respond quickly to power changes. Similarly, some modern naval designs use electric-drive architectures, modern submarines, such as the United States Navy's Columbia-class, utilize turbo-electric propulsion systems, which provide enhanced flexibility and precision in managing power demands [3–5].

Turbo-electric propulsion systems represent a transformative solution for improving the maneuverability of nuclear-powered vessels. In these systems, the reactor generates steam that drives a turbine, producing electricity. This electricity is then used to power electric motors, which drive the ship's propellers. By decoupling the reactor's power generation from the propulsion system, turbo-electric propulsion allows for smoother transitions between power levels and greater adaptability to operational needs. This configuration enables ships to maintain stable reactor output while meeting the variable demands of propulsion, significantly reducing wear and tear on reactor components. Turbo-electric propulsion has already been successfully implemented in various naval and commercial vessels, such as Russian nuclear-powered icebreakers, which rely on this technology to navigate Arctic waters with precision and reliability [5, 14].

Future designs for nuclear-powered vessels are likely to combine turbo-electric propulsion systems with advanced energy storage technologies, creating hybrid systems that offer unparalleled flexibility. For instance, during near-shore operations, energy stored in batteries or thermal reservoirs could be used to power the propulsion system, allowing the reactor to continue operating at a steady output. Offshore, the reactor could operate at full power, with excess energy stored for later use during maneuvers or emergencies. Such hybrid configurations would ensure optimal efficiency while addressing the diverse power demands of maritime operations.

Advanced reactor designs, such as molten salt reactors (MSRs) and lead-cooled fast reactors (LFRs), offer additional opportunities to improve maneuverability. These reactors are inherently more adaptable to variable power demands due to their higher thermal efficiency and operational flexibility. MSRs, for example, use liquid fuel that can be continuously circulated and reconditioned, enabling them to respond dynamically to changing power requirements. LFRs, with their high heat capacity and low-pressure operation, can maintain stable performance even under fluctuating power demands, making them particularly suitable for maritime propulsion [1–4].

In summary, the maneuverability of nuclear-powered vessels has evolved significantly from the early days of the NS Savannah to the advanced capabilities demonstrated by modern submarines and icebreakers. By integrating innovative technologies such as turbo-electric propulsion and energy storage systems, coupled with advancements in reactor design, nuclear-powered ships can achieve the flexibility needed for safe and efficient operation in both near-shore and offshore environments.

These developments not only enhance the practicality of nuclear propulsion but also pave the way for its broader adoption in commercial shipping, aligning with industry goals for sustainability and operational excellence.

3. Costs associated with nuclear reactors on ships: Analyzing historical and current data

The integration of nuclear reactors into maritime vessels has always involved significant financial considerations, encompassing capital expenditures, operational costs, regulatory compliance, and maintenance requirements. Despite the high initial costs, nuclear propulsion offers compelling long-term benefits, such as reduced fuel consumption, extended operational lifespans, and independence from volatile fuel markets. These benefits are particularly advantageous for ships operating on long-haul routes or in remote regions where refueling opportunities are limited. This section provides an in-depth analysis of the costs associated with nuclear propulsion, drawing on historical data, modern advancements, and projected future trends.

3.1 Historical insights: The NS Savannah case study

The NS Savannah, the world's first nuclear-powered merchant ship, provides an essential starting point for understanding the financial implications of nuclear propulsion. Commissioned in 1959 as a demonstration project, the Savannah cost approximately $46.9 million to construct, with $28.3 million allocated to the nuclear reactor and its initial fuel core. Adjusted for inflation, these figures translate to hundreds of millions in today's dollars, underscoring the significant investment required for nuclear propulsion [12].

Operational data from the Savannah reveals further insights into its cost dynamics. While the ship benefited from reduced fuel consumption compared to diesel-powered vessels, it faced high operational costs due to its experimental nature and the lack of supporting infrastructure. For instance, the Savannah required specialized facilities for refueling and waste management, which were prohibitively expensive given that these systems were developed for a single vessel. Furthermore, the need for a highly trained crew, coupled with stringent safety and regulatory requirements, added to its overall costs [12, 13].

The Congressional report "Costs of Operating the Nuclear Ship Savannah" highlighted that its operating expenses exceeded those of conventional ships of comparable size and capacity. However, it is important to note that these costs were influenced by the ship's status as a prototype, and they do not necessarily reflect the potential economics of a fleet of nuclear-powered vessels [12].

3.2 Modern cost considerations: The role of advanced reactors

Modern advancements in reactor technology have significantly altered the cost landscape of nuclear propulsion. Small modular reactors (SMRs) and Generation IV reactors, such as molten salt reactors (MSRs) and lead-cooled fast reactors (LFRs), are designed with cost efficiency in mind. SMRs, for instance, are factory-built and transported to the installation site, significantly reducing construction timelines and associated costs. Additionally, these reactors are smaller and require less specialized infrastructure, making them more adaptable for maritime applications [6, 7].

Fuel efficiency is another critical factor influencing the costs of nuclear propulsion. Nuclear reactors have an exceptionally high energy density, enabling ships to operate for years without refueling. For example, Russian Arktika-class nuclear-powered icebreakers utilize highly enriched uranium fuel with refueling intervals of up to 10 years. Although the upfront cost of nuclear fuel is higher than that of conventional fuels, the long operational periods between refuelings provide substantial cost savings over the vessel's lifecycle [5, 14].

3.3 Key cost drivers for nuclear-powered vessels

The overall cost structure associated with nuclear propulsion in maritime applications is shaped by a combination of capital investment, fuel-cycle characteristics, regulatory obligations, and lifecycle considerations. Unlike conventional propulsion systems, nuclear-powered vessels require substantial upfront capital expenditures related to reactor construction, containment systems, shielding, and redundant safety features. These costs reflect not only the inherent complexity of nuclear technology but also the use of specialized materials and conservative design margins necessary to ensure safe operation in a maritime environment. As a result, the initial capital cost of a nuclear-powered vessel is typically higher than that of a conventionally powered ship of comparable size and capability [5, 11].

Fuel-cycle costs represent another defining element of the economic profile of nuclear propulsion. The cost of nuclear fuel is influenced by enrichment level, fuel composition, and refueling interval. Highly enriched fuels (including HALEU and, in some naval contexts, HEU) can enable extended operational periods between refueling events in some designs, but they require more complex and costly enrichment processes. Enrichment and fabrication can account for a substantial fraction of nuclear fuel cost, and enrichment is commonly described as nearly half of nuclear fuel cost under typical assumptions [15]. Because uranium supply concentration can influence market dynamics and price volatility, major producer profiles (e.g., Kazakhstan) are often referenced in market and policy discussions [16]. In contrast, advanced reactor concepts offer alternative approaches that can reduce fuel-handling and outage-related costs over the vessel's lifetime. Molten salt reactors (MSRs), for example, employ liquid fuel that may be continuously reconditioned during operation, reducing the need for scheduled refueling outages. Lead-cooled fast reactors (LFRs) are often designed with long-lived cores that can operate for extended periods with limited refueling, thereby reducing fuel-handling logistics and associated downtime [1–4].

Regulatory compliance and safety requirements also contribute substantially to both capital and operating costs. Nuclear-powered vessels are subject to stringent international and national regulations governing radiation protection, waste management, emergency preparedness, and environmental monitoring. Compliance with these frameworks necessitates investments in specialized monitoring equipment, safety systems, documentation, and personnel training. While these requirements increase overall costs, they are essential for ensuring safe operation and public acceptance of nuclear propulsion in civilian maritime contexts [11].

Lifecycle costs further differentiate nuclear propulsion from conventional alternatives. Routine maintenance of nuclear-powered vessels includes periodic inspections, component replacement, and system upgrades to meet evolving safety and regulatory standards. In addition, end-of-life considerations, such as reactor decommissioning and the disposal or long-term storage of spent fuel and radioactive components, represent significant cost elements that must be accounted for in any comprehensive economic assessment. Advances in reactor design and waste management technologies have the potential to reduce these costs, but they remain an important factor in lifecycle planning [8, 17]. Advances in recycling and waste-management technologies may reduce waste volumes in some advanced reactor concepts; for MSR fuel/salt management and recycling pathways, see Ref. [18].

The economic viability of nuclear propulsion is therefore highly dependent on the operational profile of the vessel. Ships operating on long-haul routes or in remote and infrastructure-limited regions—such as icebreakers or transoceanic cargo vessels—are more likely to benefit from nuclear propulsion due to reduced reliance on frequent refueling and greater operational autonomy. In contrast, vessels operating on short routes with readily available fuel infrastructure may find conventional propulsion systems more economically favorable [5]. Additionally, nuclear-powered vessels benefit from relatively stable and predictable fuel costs, as nuclear fuel typically represents a smaller fraction of total operating expenses and is less susceptible to market volatility than fossil fuels. This stability can provide long-term economic advantages, particularly for operators seeking cost predictability over extended service lives [19].

Looking ahead, advancements in reactor technology are expected to further improve the economic competitiveness of nuclear propulsion. Generation IV reactor concepts, including MSRs and LFRs, incorporate features specifically aimed at reducing both capital and operational costs. Continuous fuel reconditioning in MSRs and extended core lifetimes in LFRs offer pathways to eliminate mid-life refueling, simplify logistics, and align reactor service life with that of the vessel itself. Together, these innovations suggest that future nuclear-powered ships could achieve significantly improved cost performance compared to earlier generations, such as the NS *Savannah*, while meeting modern safety and regulatory expectations [2–4].

4. Refueling costs: Historical data and modern comparisons

The NS Savannah, as the first nuclear-powered merchant ship, provides a valuable case study for understanding refueling costs in the context of early nuclear propulsion. The Savannah's reactor required enriched uranium fuel, which was expensive to produce and procure. Additionally, the refueling process itself was complex, requiring specialized facilities, equipment, and highly trained personnel. The limited availability of nuclear infrastructure at the time further compounded these costs, as only a few facilities worldwide were equipped to support civilian maritime nuclear refueling operations [12, 13]. One of the significant challenges for the Savannah was the downtime associated with refueling. Unlike conventional ships, which could refuel quickly at widely available facilities, the Savannah had to undergo a lengthy and highly regulated refueling process. This extended downtime not only incurred additional costs but also reduced the vessel's operational efficiency, highlighting the challenges of early nuclear propulsion implementations [12, 13].

Modern nuclear reactors designed for maritime applications have addressed many of the historical challenges associated with refueling through improved fuel performance, longer core life options in specialized maritime systems, and more standardized refueling/maintenance approaches. For commercial light-water power reactors, refueling is commonly described at intervals of 12, 18, or 24 months, and U.S. reactors are typically refueled every 18–24 months depending on operating strategy and outage planning [20, 21]. In contrast, open sources describing Russia's nuclear-powered icebreaker fleet discuss refueling intervals on the order of a decade for certain reactor designs, reducing the frequency of refueling outages and associated downtime and logistics burden [5, 14]. While enrichment levels can influence achievable core life in specific applications, the practical maritime advantage is reduced refueling frequency and reduced dependence on routine fuel-supply logistics relative to conventional marine fuels [5, 14].

Despite advancements, nuclear refueling still involves significant infrastructure and safety costs. Facilities for refueling must meet stringent safety standards, including radiation shielding, waste management systems, and emergency response capabilities. These requirements necessitate substantial upfront investment and ongoing maintenance expenses. Spent nuclear fuel management further adds to costs, requiring secure transport, storage, and eventual disposal in long-term repositories [8, 11, 17]. Modern solutions such as dry cask storage systems and improved waste-management techniques are widely documented in authoritative references [8, 17]. Innovative reactor designs, particularly those from the Generation IV family, are set to change nuclear refueling practices through longer service intervals and different fuel/salt concepts. Long-life LFR concepts, including sealed-core designs, have been described with "20-to-30-year-lifetime sealed core" features [22]. Historical operating and servicing costs for the nuclear-powered merchant vessel NS Savannah are summarized in **Table 1**, illustrating the contribution of nuclear-specific overhead, training, and shore-based support to overall operating expenses [12].

Cost item	**Year 1**	**Year 2**	**Year 3**
Operating costs			
Voyage expenses	1,859,283	1,948,090	1,919,613
Port charges	410,317	402,001	445,086
Cargo handling	797,441	902,095	1,122,481
Total operating costs	3,067,041	3,252,186	3,487,180
Less: voyage revenue	(2,351,702)	(2,614,421)	(2,737,530)
Net operating loss	715,339	637,765	749,650
Nuclear-related overhead and support			
Allowance for American export overhead	197,100	215,715	233,600
FAST compensation	200,000	173,260	37,478
Nuclear shore staff	305,409	441,780	480,000
Total paid to FAST	1,417,848	1,468,520	1,500,728
Nuclear servicing facility – Todd (Note a)	1,011,800	966,500	1,509,000
Training programs (Note a)	282,000	285,000	330,600
Total net cost of operating and servicing	2,711,648	2,720,020	3,340,328

Note: Parentheses denote revenue/negative values. Source: GAO B-136209 [12].

Table 1.
Operating and servicing costs (USD, nominal) for the nuclear-powered merchant vessel NS Savannah (years 1–3), including operating costs, voyage revenue, and nuclear-related overhead/support categories.

DOI: http://dx.doi.org/10.5772/intechopen.1009829

5. Operating expenses (OPEX) for nuclear-powered vessels

Operating expenses (OPEX) are one of the most significant aspects of assessing the long-term viability of nuclear-powered vessels. Unlike conventional ships, nuclear-powered vessels have unique cost structures stemming from the specialized technologies, training, maintenance, and regulatory frameworks required for their operation. This section delves deeper into the multifaceted nature of OPEX for nuclear-powered vessels, including crew training, maintenance, regulatory compliance, waste management, insurance, and energy costs, while highlighting advancements aimed at reducing these expenses.

5.1 Crew training and expertise

The operation of nuclear-powered ships requires a highly skilled crew trained in reactor management, radiation safety, emergency protocols, and maintenance procedures. The complexity of nuclear reactors demands rigorous training programs that far exceed the requirements for conventional vessel crews. The early experiences of the NS Savannah revealed the challenges of establishing such training programs, as no established standards existed at the time. Operators were trained through bespoke programs that relied heavily on naval expertise, which increased costs due to the lack of scalable solutions [12, 13].

In the modern era, the cost of crew training has been mitigated by technological advancements, such as high-fidelity simulation tools. These simulators replicate reactor operations and emergency scenarios, enabling crew members to practice in a controlled environment without exposing the actual reactor to unnecessary risks. For instance, Russia's nuclear icebreaker program employs state-of-the-art simulators to train operators, reducing training costs and improving readiness. Similarly, the U.S. Navy's nuclear propulsion training facilities have integrated virtual reality systems to enhance learning efficiency while lowering costs associated with physical training infrastructure [23, 24].

However, the challenge remains to adapt these military-grade training models to commercial nuclear-powered vessels, where the operational and regulatory requirements differ. Expanding these systems for civilian use involves additional costs, such as adapting simulation software to commercial reactor designs and ensuring compliance with maritime regulations [11].

5.2 Maintenance and lifecycle costs

Maintenance costs for nuclear-powered vessels represent a significant component of OPEX due to the complexity of reactor systems. Routine inspections, component replacements, and system upgrades are essential to maintain the safety and efficiency of the vessel. The NS Savannah faced significant maintenance challenges due to its experimental reactor design, requiring frequent inspections and adjustments. The downtime associated with these activities further increased operating costs, as the ship could not generate revenue during maintenance periods [12, 13].

Modern reactors, such as lead-cooled fast reactors (LFRs) and molten salt reactors (MSRs), are designed to minimize maintenance requirements through advanced materials and inherent safety features. For example, LFRs use liquid metal coolant, which reduces corrosion and wear on reactor components, extending their operational lifespan. Similarly, MSRs operate at atmospheric pressure, reducing the risk of mechanical failure and simplifying maintenance procedures [25]. Similarly, MSRs operate at low pressure, reducing the risk of high-pressure mechanical failure modes and simplifying certain maintenance considerations [2].

The lifecycle costs of nuclear-powered vessels also include decommissioning expenses. The safe disposal of reactor components and radioactive waste requires specialized facilities and long-term storage solutions. These costs can be substantial, and major institutional and regulatory sources document spent fuel storage options such as dry cask storage [8, 17]. For MSR fuel/salt management and recycling/reprocessing pathways that may reduce waste volumes in some concepts, see Ref. [18].

5.3 Regulatory compliance and insurance

Regulatory compliance is a cornerstone of operating nuclear-powered vessels. International frameworks, such as those established by the International Atomic Energy Agency (IAEA) and the International Maritime Organization (IMO), impose stringent requirements on radiation safety, waste management, and emergency preparedness. Compliance with these regulations involves periodic audits, safety drills, documentation and recordkeeping, and the maintenance of radiation monitoring and alarm systems, all of which contribute to OPEX [26].

Insurance costs for nuclear-powered ships are another significant factor. These vessels face higher premiums due to the perceived risks associated with nuclear technology, including potential reactor malfunctions, radiological release scenarios, and the complexities of waste handling and port interface requirements. However, advancements in reactor safety—such as passive decay-heat removal, inherent shutdown behavior, and improved containment and confinement strategies—could improve insurability and reduce premiums, particularly for vessels operating under robust regulatory frameworks with clear responsibilities, credible emergency plans, and transparent oversight [26, 27].

A useful way to think about insurance is that it reflects not only the technology, but also the operating ecosystem: training standards, operating procedures, maintenance programs, incident reporting culture, emergency preparedness coordination with ports and coastal authorities, and demonstrated performance over time. In that context, long-running nuclear maritime operations—such as Russia's nuclear icebreaker fleet—are often referenced in discussions of institutional experience with nuclear marine systems and the value of accumulated operational lessons [5]. A mature operating model with documented procedures, recurrent drills, and stable oversight can help insurers better characterize risk and may support more favorable terms than would be available for a first-of-a-kind commercial deployment.

In contrast, commercial vessels operating across multiple jurisdictions may encounter higher insurance costs because the regulatory and liability environment can vary by flag state and port state, and because insurers may price uncertainty when standards, enforcement, or emergency-response interfaces differ from one operating region to another [26, 27].

5.4 Waste management and environmental considerations

The management of radioactive waste is a unique and challenging aspect of OPEX for nuclear-powered vessels. Waste management includes the storage, transportation, and disposal of spent fuel and other radioactive materials. Early nuclear-powered ships, such as the NS Savannah, faced significant challenges in this area due to the lack of established infrastructure for maritime nuclear waste disposal [28].

Modern advancements in waste management have reduced these costs. Dry cask storage systems, for example, provide a secure and cost-effective solution for the long-term storage of spent fuel [17, 29]. Additionally, advanced recycling technologies enable the recovery of valuable isotopes from spent fuel, reducing waste volumes and creating opportunities for secondary revenue streams [30]. Environmental considerations also play a role in waste management costs. International efforts to promote sustainable nuclear practices have led to the development of stricter environmental standards for waste handling. While these standards increase OPEX, they also enhance public trust and pave the way for broader acceptance of nuclear-powered ships in the commercial sector.

5.5 Energy costs and fuel efficiency

One of the primary advantages of nuclear propulsion is its superior fuel efficiency, which can significantly reduce energy costs over a vessel's operational life by minimizing refueling frequency and fuel-supply logistics. Nuclear fuel provides unparalleled energy density, allowing ships to operate for years without refueling. For example, public sources describing Russia's nuclear icebreaker program discuss refueling intervals on the order of a decade for some designs [5, 14].

The stability of uranium markets further enhances the cost-effectiveness of nuclear propulsion. Unlike fossil fuels, which are often purchased closer to spot pricing and can be subject to rapid volatility, uranium for civilian reactors is predominantly procured through long-term contracts, supporting more predictable fuel budgeting and reducing exposure to short-term market swings [31]. This is particularly advantageous for vessels operating in remote regions, where conventional fuel delivery logistics are costly and complex [5, 14].

5.6 Future trends in OPEX reduction

Emerging technologies and reactor designs promise to further reduce OPEX for nuclear-powered vessels. Generation IV reactors, such as MSRs and LFRs, are designed with long-life cores and/or fuel-management approaches that can better align the operational lifespan of the reactor with that of the vessel. In particular, long-life sealed-core concepts have been described for some LFR designs, and MSR technology roadmaps discuss fuel/salt management approaches that can reduce outage-driven refueling requirements in certain variants. Together, these characteristics can reduce refueling frequency, simplify logistics, and potentially lower maintenance-related downtime and associated costs [18, 22].

Digitalization is also playing a growing role in OPEX reduction. The integration of advanced instrumentation and control (I&C), online monitoring, and condition-based/predictive maintenance tools enables operators to identify and address potential issues before they escalate, reducing downtime and repair costs. In parallel, digital systems and guidance are being developed to support implementation and oversight of digital I&C modernization, which can also improve the efficiency of compliance and inspection activities over time [32, 33].

While detailed operating cost data for the NS Savannah is limited, several references provide some insights into the vessel's OPEX. For example, the GAO report "Costs of Operating the Nuclear Merchant Ship Savannah" summarizes operating and servicing costs over the first 3 years (1962–1964), highlighting the contribution of nuclear-specific overhead, training, and shore-based support to total costs [12].

5.7 Key factors influencing OPEX

For any nuclear-powered vessel, the following factors will play a significant role in determining operating expenses:

Fuel costs: While nuclear fuel is generally more expensive than conventional fuels like diesel or natural gas, the cost is offset by the much higher energy density of nuclear fuel and the longer periods between refuelings.

Maintenance and repair: Nuclear reactors require regular maintenance to ensure safe and efficient operation. This includes periodic inspections, component replacements, and repairs to the reactor system and associated infrastructure.

Regulatory compliance: Operating a nuclear-powered vessel requires adherence to stringent safety and environmental regulations, which can add to the overall OPEX. These regulations cover areas such as radiation protection, emergency preparedness, and waste management.

5.8 Comparison with conventional ships

The operating expenses of a nuclear-powered vessel differ significantly from those of a conventionally powered ship. While nuclear vessels benefit from lower fuel costs over their operational lifetimes, they incur higher upfront costs and ongoing expenses related to reactor maintenance, refueling, and regulatory compliance. The trade-off between these costs will depend on the specific application of the vessel and the operational profile.

For example, nuclear-powered ships are more cost-effective for long-haul routes where refueling opportunities are limited, such as icebreakers operating in the Arctic or transoceanic cargo ships. On the other hand, conventional ships may be more economical for shorter routes or operations in areas with readily available refueling infrastructure.

5.9 Conclusion on OPEX

While nuclear-powered vessels may incur higher upfront and regulatory costs compared to conventional ships, their lower fuel costs and longer operational periods between refuelings can make them economically viable for certain applications. Detailed cost data for advanced nuclear-powered vessels are not yet available, but historical data from the NS Savannah and ongoing operations of Russian nuclear icebreakers provide valuable insights into the potential OPEX for future nuclear-powered ships.

6. Comparison of refueling intervals for PWRs and MSRs

Refueling intervals are a critical factor in the economic and operational viability of nuclear reactors, particularly for ship propulsion. Pressurized Water Reactors (PWRs), the most widely deployed reactor type, and Molten Salt Reactors (MSRs), an emerging technology family, offer contrasting approaches to refueling and fuel management. For maritime applications, the relevant metric is not only *fuel-cycle interval* but also the associated downtime, infrastructure requirements, and regulatory/port interface burden.

6.1 Pressurized water reactors (PWRs)

PWRs require periodic refueling, typically every 12, 18, or 24 months depending on design and operational strategy, and U.S. reactors are commonly refueled every 18–24 months [20, 21]. Military applications may use higher-enriched fuels (including HALEU and, in some naval contexts, HEU), allowing for longer core lives in specific mission-driven designs; however, for commercial applications, lower enrichment levels generally necessitate more frequent refueling and conventional outage-driven refueling operations.

In maritime contexts, PWR-based ship propulsion must accommodate variable power demand, and refueling strategy must account for route structure and port access. Historical examples such as the NS Savannah highlight the variability: design descriptions cite an expected operating period between refuelings of about 3.5 years [34]. In addition, long-running operational programs such as Russia's nuclear icebreaker fleet illustrate how extended refueling intervals can be achieved in specialized maritime systems; open sources describe refueling intervals on the order of a decade for some icebreaker reactor designs [5, 14].

6.2 Molten salt reactors (MSRs)

In contrast, MSRs enable alternative fuel management approaches. In some MSR variants, fuel is in a liquid salt form and may support online fuel/salt management (including removal of certain reaction products and addition of fuel constituents) without the same style of outage-driven refueling required by solid-fueled reactors [2, 18]. This can reduce the need for scheduled refueling outages in certain concepts, potentially increasing availability and reducing downtime-driven costs—though the specific benefits depend strongly on the MSR subtype, the chemical processing approach, and the maturity of the associated salt-handling systems [18].

6.3 Lead-cooled fast reactors (LFRs)

LFRs present an alternative with extended core lifespan in some concepts. Long-life sealed-core designs have been described with "20-to-30-year-lifetime sealed core" features in certain LFR concepts, aligning reactor service life more closely with vessel lifetime and reducing mid-life refueling logistics [22]. Historical experience with lead-bismuth eutectic (LBE) in ship/submarine reactor contexts has been documented in the open literature and is often referenced in discussing lessons learned for lead-cooled systems and maturation pathways [25].

6.4 Conclusion

PWRs, MSRs, and LFRs each offer distinct advantages for maritime applications. PWRs are mature and widely implemented but typically require outage-driven refueling at intervals that depend on enrichment and operating strategy [20, 21]. MSRs can offer alternative fuel management strategies that may reduce the operational penalty of refueling outages in some designs [2, 18]. LFRs include concepts aimed at long sealed-core lifetimes that could reduce refueling logistics for long-haul and remote operations [22], supported by historical experience with lead-bismuth systems as a technical foundation for ongoing development discussions [25].

7. Fluctuations in fuel rod prices and historical data

The cost of nuclear fuel (including fabricated fuel assemblies for solid-fueled reactors) is a critical element in the economic evaluation of nuclear propulsion systems. Compared with fossil fuels used in conventional shipping, nuclear fuel cost is often less exposed to short-term market swings because uranium procurement is frequently structured through longer-term contracting and fuel cost is only one part of overall nuclear operating cost. This stability, coupled with the high energy density of nuclear fuel, can be advantageous for long-term maritime applications especially where endurance and fuel logistics dominate operating costs [19, 31, 35].

7.1 Historical trends in fuel prices

Nuclear fuel costs have often been described as relatively stable compared to fossil fuels, particularly because uranium purchases for the civilian market are largely made through long-term contracts, supporting predictability in fuel budgeting and reducing exposure to near-term market swings [31]. In contrast, widely used energy price series for crude oil and natural gas show substantial variability over time, which can materially affect ship operating budgets for fossil-fueled vessels [36, 37].

A practical illustration is the operation of nuclear-powered icebreakers in Russia, where long refueling intervals and reduced dependence on routine fuel deliveries can support more predictable operating logistics in remote Arctic regions [5, 14].

7.2 Factors affecting fuel rod costs

Several variables contribute to the cost of nuclear fuel rods, ranging from raw material prices to the complexity of enrichment and fabrication. Understanding these factors is essential for evaluating the overall economic viability of nuclear-powered vessels. The cost of uranium, the primary raw material for nuclear fuel, is influenced by mining capacity, geopolitical factors, and global demand. Supply concentration and production dominance by major producers can affect market expectations and is often referenced in market and policy discussions. Kazakhstan is documented as a major uranium producer in authoritative profiles, illustrating the importance of large suppliers in uranium market dynamics [16].

Enrichment levels significantly affect fuel costs. Enrichment is a major contributor to overall nuclear fuel cost, and authoritative summaries commonly describe enrichment as accounting for a large fraction—often characterized as nearly half—of total nuclear fuel cost under typical assumptions [15, 35]. In specialized applications (including some naval contexts), higher enrichment can enable longer core lifetimes and extended operational periods between refueling events, but this comes with increased complexity and cost in enrichment capability and fuel-cycle management [15, 35].

Fuel fabrication costs depend on reactor type and fuel form, including cladding, assembly, quality assurance requirements, and any specialized materials. Advanced reactor concepts can require different fabrication pathways and materials. For MSR concepts, fuel and salt management approaches (including salt chemistry and handling of fission products) are discussed in institutional technical reports and major compilations of MSR technology status [2, 18]. For other advanced reactor families, fuel form and materials choices can similarly shift fabrication complexity and cost drivers [1, 35].

7.3 Economic impact of fuel rod costs on ship operations

The cost of nuclear fuel rods and assemblies typically represents a smaller fraction of total operating expenses for nuclear-powered ships than fossil fuel does for fossil-fueled ships, primarily because nuclear fuel has very high energy density and is procured far less frequently over a vessel's operating life. The operational advantage is therefore not only the unit cost of fuel but also reduced exposure to frequent procurement cycles and reduced reliance on recurring fuel delivery logistics, which can be significant cost drivers for conventional vessels [31, 35].

In remote or infrastructure-limited environments, the logistical costs of delivering conventional fuels can become material contributors to total operating cost. Nuclear-powered vessels avoid some of these constraints because long operating periods between refueling events reduce dependence on fuel supply chains and allow more stable long-term planning [5, 14, 31].

Russian nuclear-powered icebreakers provide a concrete operational example: long refueling intervals and reduced dependence on fuel delivery logistics can be particularly valuable in harsh Arctic environments where conventional supply chains are expensive and vulnerable [5, 14].

7.4 Fluctuations in fuel rod prices and advanced reactors

Advanced reactor technologies, including MSRs and LFRs, are designed to optimize fuel utilization and, in some concepts, reduce the operational penalty of discrete refueling outages. In MSR variants that emphasize online fuel/salt management, operational concepts may include removal of certain reaction products and addition of fuel constituents during operation, reducing reliance on outage-driven refueling in the traditional sense. The technical status, benefits, and challenges of MSR fuel/salt management are summarized in institutional reports and major technical references [2, 18].

LFR concepts can also reduce refueling-driven operational burdens in certain designs. Long-life sealed-core concepts have been described with 20–30 year lifetime sealed cores in some conceptual LFR designs, which could align reactor service life with vessel service life and reduce mid-life refueling logistics [22]. Historical and technical experience with lead-bismuth systems in ship reactor contexts has been documented and informs ongoing discussions about the maturation pathways and operational considerations for lead-cooled systems [25].

7.5 Conclusion

The stability of nuclear fuel costs, combined with the high energy density and long operating periods between refueling events, can make nuclear propulsion economically attractive for certain maritime applications, especially where endurance and fuel logistics dominate lifecycle cost and risk. Nuclear fuel cost behavior is shaped by procurement structure (often long-term contracting), enrichment and fabrication contributions to total fuel cost, and supply concentration in uranium markets [15, 16, 31, 35]. For maritime nuclear propulsion, the practical advantage is not only cost but also operational predictability and reduced logistics burden [5, 14, 31, 35]. Advanced reactor concepts may further reduce the operational impact of refueling depending on design maturity and implementation, including MSR fuel/salt management strategies and long-life LFR sealed-core concepts [18, 22].

8. Conclusions

Nuclear propulsion for merchant and offshore vessels is no longer a purely historical curiosity; it is being reconsidered in a decarbonization era where endurance, energy security, and emission constraints are becoming decisive design variables. The central lesson from early experience (especially NS Savannah) is that technical feasibility alone is insufficient—commercial viability depends on an integrated approach that couples propulsion architecture, fuel-cycle strategy, ship integration constraints (mass/volume/shielding/heat rejection), and a credible regulatory-and-insurance pathway.

From an operational standpoint, the most practical merchant concepts are those that manage highly variable ship power demand without forcing aggressive thermal cycling of the nuclear plant. Turbo-electric architectures and hybrid buffering strategies can decouple reactor operation from propulsor transients, improving maneuverability and potentially reducing maintenance-driven downtime. From an economic standpoint, nuclear propulsion tends to front-load cost through capital investment, QA/safety systems, and compliance overhead, while back-loading value through long refueling intervals, predictable fuel procurement, and reduced exposure to fuel logistics—advantages that become most compelling for long-haul routes, remote operations, and missions where refueling is costly or operationally limiting.

Advanced reactor concepts broaden the design space, but they do not remove the need for careful fuel-cycle and lifecycle planning. MSR and LFR concepts introduce alternative fuel/salt management and long-life core strategies that can reduce refueling-driven outages in certain variants; however, these benefits must be evaluated within an end-to-end safety case and a complete lifecycle plan that includes decommissioning and waste pathways. Ultimately, merchant nuclearization is best framed as a systems-engineering and evidence-package problem: the path to adoption will depend on demonstrating safe, maintainable, and insurable operation across jurisdictions—supported by standardized procedures, training, drills, monitoring, and transparent oversight.

Acknowledgements

The authors gratefully acknowledge the support and technical engagement of the American Bureau of Shipping (ABS). ABS support helped motivate a rigorous, engineering-focused treatment of nuclear power as a prospective maritime propulsion option, with emphasis on safety, operational practicality, and the evidence needed for credible stakeholder acceptance. The views expressed in this chapter are those of the authors and do not necessarily represent the official views, policies, or positions of ABS.

Author details

Nima Fathi[1]* and Milton Korn[2]

1 Texas A&M University, TX, USA

2 Texas A&M Maritime Academy, TX, USA

*Address all correspondence to: nfathi@tamu.edu

References

[1] Advanced Reactors Information System (ARIS). International Atomic Energy Agency (IAEA). Available from: https://www.iaea.org/resources/databases/advanced-reactors-information-system-aris

[2] Molten Salt Reactors. World Nuclear Association. Available from: https://world-nuclear.org/information-library/nuclear-power-reactors/other/molten-salt-reactors

[3] Generation IV International Forum. GIF Portal. Available from: https://www.gen-4.org/

[4] Generation IV International Forum (GIF) Updates. OECD Nuclear Energy Agency (NEA). Available from: https://www.oecd-nea.org/jcms/pl_50603/generation-iv-international-forum-gif-updates

[5] Nuclear-Powered Ships. World Nuclear Association. Available from: https://world-nuclear.org/information-library/non-power-nuclear-applications/transport/nuclear-powered-ships

[6] Benefits of Small Modular Reactors (SMRs). U.S. DOE Office of Nuclear Energy. Available from: https://www.energy.gov/ne/benefits-small-modular-reactors-smrs

[7] Advanced Small Modular Reactors (SMRs). U.S. DOE Office of Nuclear Energy. Available from: https://www.energy.gov/ne/advanced-small-modular-reactors-smrs

[8] Storage and Disposal of Radioactive Waste. World Nuclear Association. Available from: https://world-nuclear.org/information-library/nuclear-fuel-cycle/nuclear-waste/storage-and-disposal-of-radioactive-waste

[9] 2023 IMO Strategy on Reduction of GHG Emissions from Ships. International Maritime Organization (IMO). Available from: https://www.imo.org/en/ourwork/environment/pages/2023-imo-strategy-on-reduction-of-ghg-emissions-from-ships.aspx

[10] MEPC.377(80). International Maritime Organization (IMO). 2023. Available from: https://wwwcdn.imo.org/localresources/en/KnowledgeCentre/IndexofIMOResolutions/MEPCDocuments/MEPC.377%2880%29.pdf

[11] Safety Standards. International Atomic Energy Agency (IAEA). Available from: https://www.iaea.org/resources/safety-standards

[12] Costs of Operating the Nuclear Merchant Ship Savannah (B-136209). U.S. General Accounting Office (GAO); 1970. Available from: https://www.gao.gov/assets/b-136209.pdf

[13] N.S. Savannah Document Library. United States Maritime Administration (MARAD). Available from: https://www.maritime.dot.gov/national-defense-reserve-fleet/ns-savannah-program/ns-savannah-document-library

[14] Bukharin O. Russia's nuclear icebreaker Fleet. Science and Global Security. 2006;**14**:25-31. Available from: https://scienceandglobalsecurity.org/archive/sgs14bukharin.pdf

[15] Uranium Enrichment. World Nuclear Association. Available from: https://world-nuclear.org/information-library/nuclear-fuel-cycle/

conversion-enrichment-and-fabrication/uranium-enrichment

[16] Uranium and Nuclear Power in Kazakhstan. World Nuclear Association. Available from: https://world-nuclear.org/information-library/country-profiles/countries-g-n/kazakhstan

[17] Dry Cask Storage. United States Nuclear Regulatory Commission (NRC). Available from: https://www.nrc.gov/waste/spent-fuel-storage/dry-cask-storage

[18] Status of Molten Salt Reactor Technology (STI-DOC-010-489). International Atomic Energy Agency (IAEA); 2023. Available from: https://www-pub.iaea.org/MTCD/Publications/PDF/STI-DOC-010-489_web.pdf

[19] Nuclear Explained: Data and Statistics. U.S. Energy Information Administration (EIA). Available from: https://www.eia.gov/energyexplained/nuclear/data-and-statistics.php

[20] Nuclear Power Reactors. World Nuclear Association. Available from: https://world-nuclear.org/information-library/nuclear-power-reactors/overview/nuclear-power-reactors

[21] Today in Energy: Status of U.S. Nuclear Outages. U.S. Energy Information Administration (EIA). Available from: https://www.eia.gov/todayinenergy/detail.php?id=60942

[22] Cinotti L et al. LFR "Lead-cooled fast reactor". In: FISA 2006. Kirchberg (Luxembourg). Available from: https://www.osti.gov/servlets/purl/889453

[23] Naval Sea Systems Command (NAVSEA). Nuclear Power School. Available from: https://www.navsea.navy.mil/Home/NNPTC/Academics/Nuclear-Power-School/

[24] Naval Nuclear Propulsion Program. Naval Nuclear Laboratory. Available from: https://navalnuclearlab.energy.gov/nuclear-propulsion-program/

[25] Zrodnikov AV et al. Use of Russian Technology of Ship Reactors with Lead-Bismuth Coolant in Nuclear Power2000. pp. 127-155. Available from: https://www.osti.gov/etdeweb/servlets/purl/20114853

[26] Requirements for Nuclear Power Systems for Marine and Offshore Applications. American Bureau of Shipping (ABS); 2024. Available from: https://ww2.eagle.org/content/dam/eagle/rules-and-guides/current/special_service/346-requirements-for-nuclear-power-systems-for-marine-and-offshore-applications-2024/346-nuclear-power-systems-reqts-oct24.pdf

[27] Klimczak J. Nuclear Opportunity and Challenges for the Shipping Industry. Available from: https://axaxl.com/fast-fast-forward/articles/nuclear-opportunity-and-challenges-for-the-shipping-industry

[28] Koehler E. Lessons learned N.S. SAVANNAH decommissioning. In: USNRC – NEI Decommissioning Workshop. 2025. Available from: https://www.nrc.gov/docs/ML2501/ML25017A037.pdf

[29] Backgrounder on Dry Cask Storage of Spent Nuclear Fuel. United States Nuclear Regulatory Commission (NRC). Available from: https://www.nrc.gov/reading-rm/doc collections/fact-sheets/dry-cask-storage

[30] Spent Fuel Reprocessing Options (IAEA-TECDOC-1587). International Atomic Energy Agency (IAEA); 2008. Available from: https://www-pub.iaea.org/MTCD/Publications/PDF/TE_1587_web.pdf

[31] Uranium Marketing Annual Report. U.S. Energy Information Administration (EIA). Available from: https://www.eia.gov/uranium/marketing/

[32] Implementing Digital Instrumentation and Control Systems in the Modernization of Nuclear Power Plants. Vienna: International Atomic Energy Agency (IAEA); 2009. Available from: https://www-pub.iaea.org/MTCD/Publications/PDF/Pub1383_web.pdf

[33] Digital Transformation: Opportunities and Challenges for the Nuclear Sector. OECD Nuclear Energy Agency (NEA); 2021. Available from: https://www.oecd-nea.org/jcms/pl_59100/digital-transformation-opportunities-and-challenges-for-the-nuclear-sector

[34] NS Savannah Propulsion Plant. Maritime.org. Available from: https://maritime.org/tour/savannah/press/part2b.php

[35] Economics of Nuclear Power. World Nuclear Association. Available from: https://world-nuclear.org/information-library/economic-aspects/economics-of-nuclear-power

[36] Europe Brent Spot Price. U.S. Energy Information Administration (EIA). Available from: https://www.eia.gov/dnav/pet/hist/rbrteD.htm

[37] Henry Hub Natural Gas Spot Price. U.S. Energy Information Administration (EIA). Available from: https://www.eia.gov/dnav/ng/hist/rngwhhdD.htm

Section 3

Technologies beyond Fission

Chapter 5

Advancements and Challenges in Nuclear Fusion Reactors Technology: A Comprehensive Overview

Amir Zacarias Mesquita, Welen Nunes de Lima,
Leonardo Yago Zwetsch Varella and Walewska Gouveia Silva

Abstract

Until now, the energy produced in nuclear reactors has come from the fission of heavy nuclei, particularly the isotope U-235. However, as is well known, the conversion of mass into energy also occurs with light nuclei. When hydrogen and deuterium fuse to form a heavier nucleus, such as tritium and helium, they release energy. Stars are the largest fusion reactor power plants. A star is initially just a cloud of hydrogen. The gravitational attraction brings hydrogen atoms together, increasing pressure, density, and temperature. Kinetic energy causes collisions to the point where electrons are separated. The mass of nuclei and electrons forms plasma, which is the fourth state of matter. Hot plasma from nuclei meets the conditions for the initiation of fusion reactions. Two techniques have been developed to enable energy production in fusion reactors. The oldest (started in the mid-1950s) is magnetic confinement, in which plasmas at thermonuclear temperatures are confined by appropriate magnetic fields. The latest technique for performing fusion (begun in the late 1960s) is inertial confinement, in which tiny solid targets are compressed to very high densities using laser beams. This chapter briefly recalls the fundamental concepts of the energy released by nuclear fusion.

Keywords: nuclear fusion, plasma, nuclear force, hydrogen, ITER®, laser, nucleus

1. Introduction

Fusion reactions of the nuclei of light atoms release an amount of energy more than a million times greater than a typical chemical reaction. This enormous amount of energy occurs because the mass of the nucleus produced is less than the sum of the masses of the initial nuclei. The lost mass is converted into energy. Although fusion is an energetically favorable (exothermic) process for light nuclei, it does not occur naturally here on Earth, due to the natural difficulties of bringing reactants together (due to electrostatic repulsion between the two nuclei) so that nuclear forces can act.

Fusion reactions have been happening for billions of years in the universe. In fact, fusion reactions are responsible for producing energy in most stars, including the Sun. Scientists on Earth have only been able to produce nuclear fusion reactions in the last 60 years. Fusion between heavier nuclei is routinely produced in small quantities in particle accelerators. Nuclear fusion is the basis of our lives since the solar energy produced by this process is essential for maintaining life on Earth [1].

When a star is created, it consists of hydrogen and helium generated in the Big Bang, the process that gave rise to the universe. Due to the enormous gravitational field, hydrogen atoms in the star collide and fuse to form helium nuclei. Later, helium, colliding with hydrogen and other helium nuclei, gives rise to heavier elements. These reactions continue until iron is formed. From Fe onward, fusion no longer occurs in the star. The process becomes energetically unfavorable. When a star has converted an appreciable fraction of its hydrogen and helium into heavier elements, it moves into the final stage of its life. Some stars begin to contract, into a sphere made up largely of iron. However, if the star's mass is large enough, a tremendous, violent, and bright explosion can occur. The star suddenly expands and produces, in a short time, more energy than the sun will produce in its entire lifetime. When this occurs, the star becomes a supernova. When the star is in the supernova phase, many important nuclear reactions take place. In the explosion, the nuclei are accelerated to speeds much greater than those they normally have in the star. In the new condition, the high-speed nuclei collide and can now fuse, producing elements with a mass greater than that of iron. The extra energy from the explosion is needed to overcome the enormous repulsive force between the nuclei due to the nuclear electrical charge. Elements such as lead, gold, and silver found on Earth were once leftovers from a supernova explosion. The iron that we find on much of the Earth's surface, as well as in its core, comes from both the remains of supernovae and dead stars [1].

When a gas is heated to several thousand degrees, the particles separate into parts (molecules dividing or atoms losing electrons, that is, becoming ionized), giving rise to the state of matter called plasma. The study of plasma is of interest for carrying out controlled thermonuclear fusion. It would be a very abundant source of energy and much cleaner than that of nuclear fission, used in common nuclear power plants. It is said to be "controlled" in contrast to the energy generated in the explosions of hydrogen bombs (H bombs). The technical and theoretical problems involved in carrying out controlled fusion are immense. Research on the subject has been going on for 60 years and it may be another 60 years before the first controlled fusion plant begins to produce electricity commercially. Large fusion experiments, based on systems called tokamaks, are being built. The most promising is ITER®, in the south of France, which involves the collaboration of several countries (European Union, Japan, USA, Russia, China, India, and South Korea) [2].

As a plasma suitable for fusion would be very hot—tens of millions of degrees Celsius, it cannot be contained normally: contact with the material walls must be avoided (this, in addition to damaging them, would also cool the plasma). To do this, you can take advantage of the fact that it is basically made up of electrically charged particles (because, due to the high temperature, the atoms would lose their electrons, becoming electrically positive). Thus, magnetic fields and electric currents induced in the plasma are used to conveniently divert the trajectories of the particles (which normally move quickly and randomly back and forth), to always keep them within a region of space. This is called "*magnetic confinement*" [2].

Working with the theory of stimulated radiation emission formulated by Einstein [3], the American physicist and engineer Theodore Harold Maiman was the first to build a Laser (Light Amplification by Stimulated Emission of Radiation) on May 16, 1960. Maiman's [4] experiment was based on the work of other pioneers such as Gordon Gould, Charles Townes, and Arthur Schawlow. Maiman [4] shined a high-power light on a ruby partially covered with silver and the result was the world's first laser.

Currently, after more than sixty years, there are thousands of applications of lasers, both for basic science and in use directly for the benefit of humanity. In December 2022, the experimental confirmation of one of the applications, which has been attempted for more than fifty years, was announced, that is, the production with the gain of fusion energy through the *inertial confinement* scheme. In the experiment carried out by scientists from the National Ignition Facility—NIF, at Lawrence Livermore National Laboratory—LLNL, the so-called ignition condition was reached, in which the energy released by the fusion reactions between deuterium and tritium supplants the energy provided by the laser beam used to initiate reactions [5].

2. Nucleon binding energy

The use of nuclear energy is possible due to the so-called mass "deficits" of the nucleus, that is, the mass of the compound atoms is not equal to the sum of the masses of the components involved. For example, in the formation of the helium nucleus, energy is released by the fusion of hydrogen atoms. On the other hand, the fission of uranium, forming two lighter elements, releases energy. In both cases, a small portion of the mass is transformed into energy, as given by Einstein's famous equation:

$$E = mc^2 \tag{1}$$

Where m is the mass loss and c is the speed of light in a vacuum (3×10^8 m/s).

In **Figure 1**, a graph is presented showing the energy released per nucleon, as a function of the mass number (A). This graph is normally shown inversely, that is, it shows the binding energy per nucleon, which is the energy required to separate a

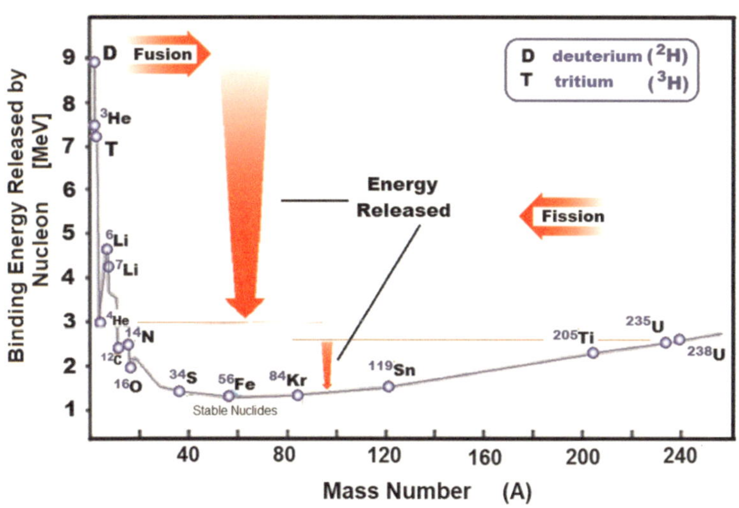

Figure 1.
Binding energy released by nucleon. Source: Mesquita [6].

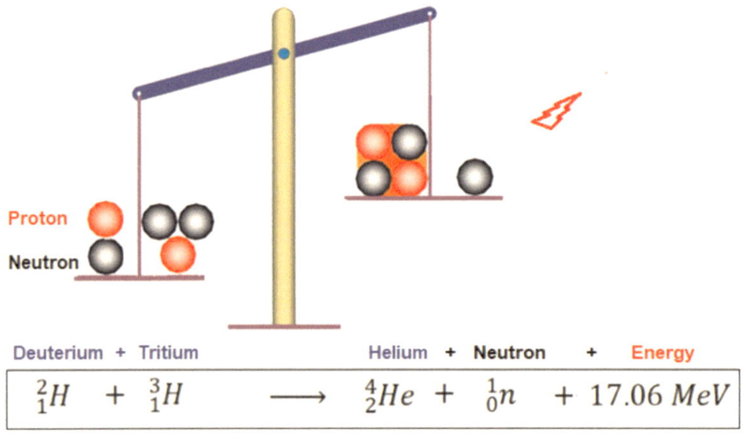

Figure 2.
Mass transformed into energy in the formation of helium. Source: Mesquita [6].

neutron or a proton from the nucleus. For light nucleons, the binding energies per nucleon show an increase until reaching a maximum value of 8.7 MeV for nuclides with approximately 50 or 60 nucleons (iron and nickel), it then gradually decreases until reaching 7.5. MeV for uranium-238. The decrease is more pronounced from $A > 150$ because of the Coulomb repulsion between the protons. This repulsion, which is small compared to the nuclear force, becomes relatively more important for nucleons with high Z values. This is why there is a limit to the number of stable elements that can be formed.

Nuclei that have higher binding energy per nucleon are more stable. In the curve shown in **Figure 1**, there are two ways in which energy can be released by atomic nuclei, namely, fusion of light nuclei or fission of heavy nuclei. In both processes, the new nuclides formed have smaller masses than the original nuclide. This difference in mass is transformed into energy (Eq. (1)).

The fusion of hydrogen forming helium is the source of energy for stars like our Sun. The fission of uranium-235 is used in nuclear reactors to produce heat and electricity. Most forms of energy used by man originate from the Sun, either as fossil fuels or even with the direct use of solar energy. The big exception is nuclear energy.[1]

An example of fusion is the formation of helium, which is lighter than the set of its components: a proton, a neutron, and H-3 (tritium or T), as illustrated in **Figure 2**. Equivalent process occurs in the splitting of heavy elements (fission) when a smaller amount of mass is transformed into energy.

3. Fusion reactions

An example of a fusion reaction is the combination of two hydrogen nuclei with two neutrons to form helium, as shown below:

---

[1] Geothermal energy also does not come from the sun but from the heat of the Earth's core, including the decay of radioactive elements present there, mainly: U-238, Th-232, and K-40. Other types of energy, used in reduced quantities, also do not originate from the Sun, such as tidal energy and thermoelectric generators.

$$2_1^1H + 2_0^1n \rightarrow {}_2^4He \tag{2}$$

Using the differences in atomic mass, we have:

$$2(1.007825) + 2(1.008665)\text{–}4.002603 = 0.030377\text{ amu} \tag{3}$$

This mass value (0.030377 amu) corresponds to 28.3 MeV of energy.

The energy released in the fusion of deuterium with two neutrons, per unit of atomic mass (28.3 MeV/4), is around 7 MeV. It is around 9 times greater than the energy of fission of uranium-235 per unit of atomic mass (200 MeV/235), which is equal to 0.8 MeV.

Several reactions are possible by combining hydrogen nuclei (${}_1^1H$) and their isotopes, deuterium (${}_1^2$H or D) and tritium (${}_1^3$H or T), or using other light elements such as lithium (Li), boron (B), and helium (He), always releasing energy. A very promising fusion reaction uses deuterium ${}_1^2$H and the isotope helium-3, as follows:

$${}_1^2H + {}_2^3\text{He} \rightarrow {}_1^1\text{H} + {}_2^4\text{He} + 18.3\text{ MeV} \tag{4}$$

Different approaches and experiments to achieve fusion with energy gains have been in evidence in several countries in recent years. The two best-known processes in research into the fusion of the nuclei of light chemical elements are presented below.

4. Magnetic confinement of plasma

The conditions to trigger fusion chain reactions are very high temperatures and pressure. These conditions have been achieved, since 1949, with the hydrogen bomb, or thermonuclear bomb (uncontrolled fusion), which is initiated with the detonation of a fission device. Controlling the fusion reaction has been attempted for several years, but, to date, it has not been possible to maintain the chain reactions in such a way that there is a positive gain in energy. In other words, more energy is spent to create and maintain the plasma confined than the energy that can be used. Plasma is the fourth state of matter, the other three being solid, liquid, and gas. It is in the plasma that fusion reactions take place.

The most common way to confine the plasma has been through a magnetic field provided by powerful electromagnets, which produce a toroidal magnetic field. Plasma is formed by the isotopes of hydrogen, deuterium, and tritium. The magnetic field limits the space where fusion occurs, so as not to touch the walls of the reactor vessel, otherwise the walls would be damaged, in addition to cooling the plasma. Plasma particles normally move in all directions due to their very high kinetic energy. As they are electrically charged, the magnetic field directs them around the center of the toroid. Positive particles rotate in one direction and negative particles (electrons) in the opposite direction. They do not leave the magnetic field lines of force and do not reach the walls. Thus, collisions eventually occur, generating fusion energy. Neutrons formed in fusion reactions carry energy to walls and heat removal systems. One of the difficulties with fusion reactors is that the nuclear reaction product needs to be removed from the system, as it is an impurity.

Almost all experimental research at present is based on magnetic confinement using the controlled fusion reaction of deuterium and tritium (D-T). Because, it demands the highest cross section, at the lowest plasma temperature to overcome electrostatic repulsion, according to the reaction:

$$^{2}_{1}H + ^{3}_{1}H \rightarrow ^{4}_{2}He + ^{1}_{0}n + 17.06 \text{ MeV} \tag{5}$$

Deuterium is stable and has an average abundance of 0.015% of the hydrogen content of water on our planet ($^{1}_{1}H$ has an abundance of 99.98%). Tritium is radioactive with a half-life of 12.3 years, decaying into He-3, and emitting beta (β) particles. Its abundance is scarce and is obtained by neutron activation reaction of the isotope lithium-6 and lithium-7,[2] with neutrons in a nuclear research reactor or using the neutrons generated in Eq. (5), according to the reactions (Eqs. (6) and (7)):

$$^{1}_{0}n + ^{6}_{3}Li \rightarrow ^{4}_{2}He + ^{3}_{1}H + 4.78 \text{ MeV} \tag{6}$$

$$^{1}_{0}n + ^{7}_{3}Li \rightarrow ^{4}_{2}He + ^{3}_{1}H + ^{1}_{0}n - 2.74 \text{ MeV} \tag{7}$$

The first assembly by magnetic confinement went into operation in 1956 at the Kurchatov Institute in Moscow, being known as Tokamak[3] (***TO**roidal'naya **KA**mera s **MA**gnitnymi **K**atushkami*) "toroidal chamber with magnetic coils". It was devised by Soviet physicists. From that time until today, it has been perfected, but the same principle has always been maintained. There are dozens of tokamaks distributed across several countries.

A variation of magnetic confinement is the stellarator. The most advanced currently is at the Max Plank Institute in Germany called Wendelstein 7-X (W7-X). Tokamaks are subject to magnetic disturbances that can destabilize the entire reactor. Differences in the way magnetic fields are created make stellarator immune to these problems. However, stellarators are much more complex to build. While tokamaks are better at keeping plasmas hot, stellarators are better at keeping them stable. Despite the current prevalence of the tokamak, it is still possible that stellarators could one day become the preferred option for a future fusion power plant. About 60 tokamaks and 10 stellarators are currently operating [7].

5. The ITER® organization: the way

Currently, the most advanced Tokamak under construction is ITER®, "The Way" in Latin (previously an acronym for International Thermonuclear Experimental Reactor), located in Caradache, in the south of France. Its construction began in 2013 and is expected to be completed in 2025 when its commissioning and plasma experiments will begin. Full deuterium-tritium (D-T) fusion experiments will begin in 2035. The project is an international cooperation involving the European Union, China, India, Japan, South Korea, Russia, and the United States, with the support of the International Atomic Energy Agency (IAEA). It will be the largest assembly of its kind in the

---

[2] Li-6 and Li-7 are stable and have abundances of 7.5 and 92.5%, respectively. It is a material found in considerable quantities in the earth's crust and seawater.

[3] Acronym of the Russian term: **то**роидальная **ка**мера с **ма**гнитными **ка**тушками.

world, twice the size and ten times the volume of the plasma chamber of the largest machine currently in operation.

Its thermal power will be 500 MW, 10 times greater than the power to be consumed by the system, which will be 50 MW. Producing a positive gain in energy is the objective of the installation. The conversion of nuclear energy into electricity is not the major objective of the project. The plasma temperature will reach $150 \times 10^{6\circ}$ C. The plasma radius will be 6.2 m and its volume 840 m^3. The magnet coils will be made of a niobium and titanium alloy weighing 10 tons. They will be cooled to a temperature of 4 K (−269°C) in supercritical helium, becoming superconductors. There will be 18 "D"-shaped toroidal field electromagnets, placed around the vessel producing the magnetic field. The main function of the vessel is to confine plasma particles. The ITER® experiments will take place inside this hermetically sealed, vacuum-sealed steel vessel, which houses the D-T fusion reactions and acts as the first safety containment barrier. In its tire-shaped chamber (torus), plasma particles spiral continuously without touching the walls. **Figure 3** shows a diagram of the ITER® tokamak. An image of a person is shown to highlight the dimensions of the system.

As can be seen, inside the ITER® tokamak, there will be a large temperature difference over a small distance, that is, 150 million °C in the plasma and almost zero kelvin (K) in the superconductors. The electrical conductors of the coils will be made of an alloy containing niobium. It is noteworthy that Brazil concentrates approximately 92% of production and more than 98% of global niobium reserves, with the largest reserves located in the state of Minas Gerais [9].

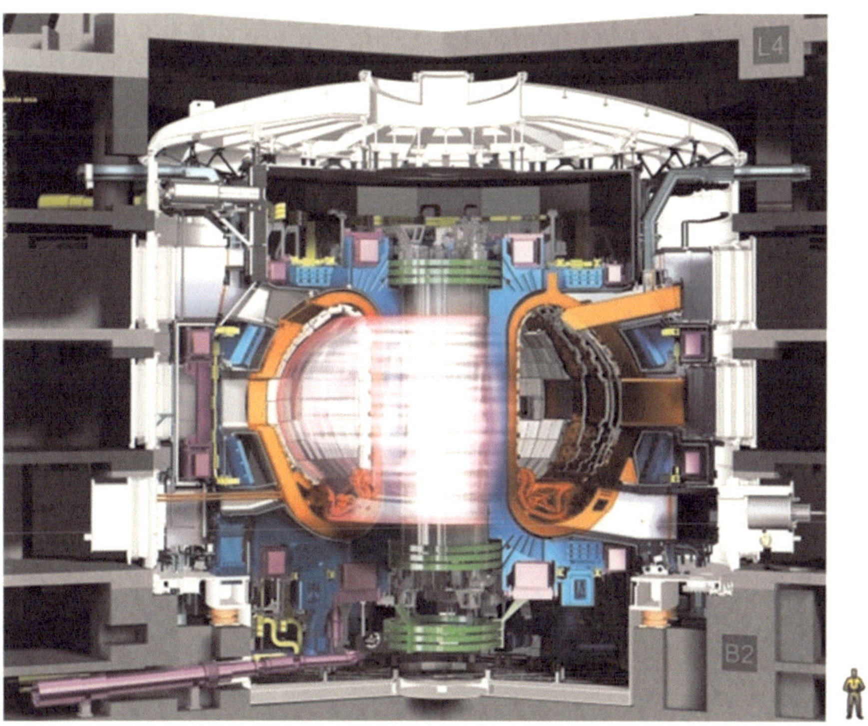

Figure 3.
Sectional drawing of the ITER® tokamak. Courtesy: ITER® Organization [8].

6. Inertial confinement: laser fusion

The United States Department of Energy (DOE) and the DOE National Nuclear Security Administration announced on December 13, 2022 that researchers at the National Ignition Facility (NIF) located at Lawrence Livermore National Laboratory (LLNL) in California have achieved more energy than laser beams used to initiate nuclear fusion reactions. Laser fusion experiments have been attempted for decades to achieve high levels of temperature and pressure to fuse nuclei. However, no nuclear fusion experiment had been able to produce more energy than the laser energy used in the experiments (laser fusion) [10].

A powerful 192-beam laser entered the top and bottom of a small metal cylinder. Inside was a plastic capsule the size of a small pea (about 4.6 mm in diameter) containing 150 micrograms of the two isotopes of hydrogen (fuel). The sudden heating of the cylinder generated an internal X-ray attack that compressed the deuterium and tritium (DT or H-2 and H-3) fuel pellet. In a moment, lasting less than 100 trillionths of a second, it uniformly increased the surface temperature of the capsule to 100 million degrees Celsius (°C), about 10 times what stars reach in fusion. The 2.05 megajoules (MJ) of laser energy inside the cylinder, equivalent to about half a kilogram of TNT, bombarded the DT pellet. This heat caused the capsule to explode and generated a shock wave that compressed the mixture of deuterium and tritium and caused nuclear fusion. Thus, neutrons and alpha particles (He nucleus) were released. The neutrons escape, but the alpha particles deposit their energy into the dense fuel, heating it further. 3.15 MJ of energy was released from the fusion reaction, a factor of 1.54 gain in relation to the energy input [11, 12].

This passed the threshold that laser fusion scientists call ignition, the dividing line where the energy generated by fusion is equal to the energy of the incoming lasers that initiate the reaction [13]. What has not been fully disclosed is that to produce the laser beam, around 150 times more electrical energy was used than was produced in the process (**Figure 4**).

Although a significantly greater energy gain is needed to produce energy, the experiment represented a huge step forward in fusion research. The laser is so strong that it can heat the capsule to 100 million °C. The temperature is higher than that of the Sun and equivalent to the temperature that would occur if the Earth's atmosphere were compressed 100 billion times. Under these forces, the capsule begins to implode in on itself, forcing the hydrogen atoms to fuse, releasing energy.

Although the experiment produced a net gain in energy compared to the energy of 2.05 MJ from the incoming laser beams, 300 MJ from the mains was required to energize the laser to generate the brief pulse. The laser used is the most powerful in the world, but it is slow and inefficient, relying on decades-old technology. The device, the size of a sports stadium, was designed to perform basic scientific experiments, not to serve as a prototype for generating electricity. He performs an average of 10 laser pulses per week. A commercial installation using the laser fusion approach would need much faster lasers, capable of firing at the rate of a machine gun, perhaps 10 times per second. The laboratory still consumes much more energy than that produced by fusion reactions. Other types of lasers are more efficient. The researchers calculate that a viable laser fusion power plant would likely require energy gains much greater than the 1.5 observed in this recent fusion experiment. Gains of 30–100 would be needed to make energy supply viable in a plant LNLL [15].

Figure 4.
The National Ignition Facility (NIF) target chamber. Source: LLNL [14].

7. Conclusion

Fusion reactions constitute the source of energy for the largest power plants in the universe: the stars. Although a star is initially just a cloud of hydrogen, contraction, caused by its own gravitational pull, increases pressure, density, and temperature. The collisions between atoms increase in number and violence until they begin to release their electrons, forming plasma. Stars carry out this process spontaneously. The effect of its gravitational field is so enormous that, naturally, it promotes extreme compression of the nuclei of these two isotopes of hydrogen, creating ideal conditions for nuclear fusion to occur.

Two schemes are used to enable energy generation in nuclear fusion reactors. The oldest system began to be researched in the 1950s and is known as magnetic confinement. In this system, plasmas reach very high temperatures, being confined by magnetic fields in a toroid-shaped reactor. These are the well-known tokamaks developed in the former Soviet Union [5].

The most recent technique, which began in the 1960s, is inertial confinement fusion. It uses laser beams that fire for a few billionths of a second to generate X-rays. The X-rays compress and heat a tiny fuel capsule (deuterium and tritium) the size of a peppercorn pea (about 4.6 mm in diameter). Subsequently, the fuel becomes hot and dense enough to form a plasma. The hydrogen nuclei begin to collide, fusing and releasing energy. It was in this type of experiment that in 2022 at the Lawrence Livermore National Laboratory (LLNL), more energy was obtained than the energy of the laser beams used to initiate nuclear fusion reactions [16].

While both techniques aim to create plasma hot enough to fuel fusion, the physics and engineering required to get there differ in several respects.

However, controlled nuclear fusion is still in the research phase and it could take decades before electricity is produced through this way. However, the prospects are attractive due to the abundance of deuterium in nature. The main project that aims to demonstrate the commercial viability of using controlled nuclear fusion through magnetic plasma confinement is ITER®, based on tokamak technology.

A nuclear fusion process that was in the news a lot recently was a fusion with energy gain using a laser beam. The experiment was carried out by researchers at Lawrence Livermore National Laboratory (LLNL) in California. However, it is very likely that the laser beam fusion process will never be used to supply electrical energy. This is a very expensive and inefficient way to cause nuclear fusion. The objective of the experiment carried out was to demonstrate that it is possible to obtain energy gains with fusion using laser beams.

Researchers at other institutes are analyzing variations of the experiment carried out. Other types of lasers at different wavelengths can heat hydrogen more efficiently. Some researchers suggest using "direct drive" for laser fusion, using laser light to directly heat hydrogen.[4] This would put more energy into the hydrogen, but it could also create instabilities that would impede fusion reactions.

Critics of research to develop fusion reactors often belittle efforts in this area with the facetious comment that "*fusion reactors are always thirty years in the future*". This scientifically retrograde attitude is worth countering with the statement of Dr. Arati Prabhakar, Chief Advisor for Science and Technology to the President of the United States of America when commenting on the NIF result, "*We have had a theoretical understanding of fusion for over a century, but the journey from knowing to doing can be long and arduous. Today's milestone shows what we can do with perseverance*" [5].

Germany was one of the main countries to develop and use fission nuclear power plants. However, its use for generating electrical energy was abolished after the Fukushima accident. Now, in 2024, Germany's Federal Minister for Research Bettina Stark-Watzinger has announced a new funding program called "*Fusion 2040*". This is an ambitious funding program for research into nuclear fusion. The objective is to pave the way for the construction of the country's first nuclear fusion plant by 2040. The program will have two phases: the first for the development of technologies, components, and materials needed for a fusion plant by the beginning of 2030. In the second phase, the focus will be on the integration of a central project, explains the minister. Together with the resources already allocated to research institutions, the German government intends to make more than one billion euros available for nuclear fusion research by 2028, involving universities, research institutes, and industry [17, 18].

Light nuclear fusion experiments are important and will be the future of nuclear energy without waste. But it will take a few decades to achieve this. Until then, the only clean and sustainable energy available, which humanity has dominated and controlled since the first nuclear reactor in 1942, is uranium fission, which will continue to dominate for several decades.

Acknowledgements

The authors would like to thank the following institutions: Nuclear Technology Development Center (CDTN), Brazilian Nuclear Energy Commission (Cnen), Research Support Foundation of the State of Minas Gerais (Fapemig), and Brazilian Council for Scientific and Technological Development (CNPq).

[4] Without transforming the laser beam into an X-ray for attack inside the cylinder.

Author details

Amir Zacarias Mesquita*, Welen Nunes de Lima, Leonardo Yago Zwetsch Varella and Walewska Gouveia Silva
Nuclear Technology Development Center/Brazilian Nuclear Energy Commission (CDTN/Cnen), Belo Horizonte, MG, Brazil

*Address all correspondence to: amir@cdtn.br

References

[1] Berkeley Lab. ABC's of Nuclear Science: Fusion. Nuclear Science Division. Lawrence Berkeley National Laboratory. California (CA), USA: University of California; 2017

[2] Machida M, Sakanaka PH. The Plasma Confinement. Group of Plasma Physics and Controlled Thermonuclear Fusion. Gleb Wataghin Institute of Physics. Campinas State University (Unicamp); 2024. Available from: https://portal.ifi.unicamp.br/en/deq/gfpftc

[3] Einstein A. The Quantum theory of radiation. Physikalische Zeitschrift. 1917; **18**:121

[4] Maiman TH. Stimulated optical radiation in ruby. Nature. 1960;**187** (4736):493-494

[5] Galvão R. About the Recent Laser Fusion Result Obtained at Lawrence Livermore National Laboratory. Sociedade Brasileira de Física. 2022. Available at: https://www.sbfisica.org.br/v1/sbf/sobre-o-recente-resultado-de-fusao-por-laser-obtido-no-lawrence-livermore-national-laboratory/. [Accessed: February 24, 2024]. (in Portuguese)

[6] Mesquita AZ. Nuclear Energy: An introduction. Ed. Universidade Federal do Paraná (UFPR). 1st ed. Vol. 1. Curitiba (PR), Brazil. 2024. 249 p. ISBN: 9788584802210. (in Portuguese)

[7] IAEA—International Atomic Energy Agency. Fusion Energy. Vienna: IAEA Bulletin; 2021. pp 6-7

[8] ITER Organization. The ITER Tokamak. 2024. Available from: https://www.iter.org/mach

[9] Pestana A. ITER: The Paths of Fusion Energy and Brazil. Brasília: Funag; 2015. (In Portuguese)

[10] US DOE. Department of Energy. DOE national laboratory makes history by achieving fusion ignition. 2024. Available from: https://www.energy.gov/articles/doe-national-laboratorymakes-history-achieving-fusion-ignition

[11] Pivetta M. Nuclear fusion experiment produces more energy than it consumes for the first time. Revista Pesquisa Fapesp. Ed. Fundação de Pesquisa do Estado de São Paulo. 2023; **324**:39-42

[12] Tollefson J, Gibney E. Nuclear-fusion lab achieves 'ignition': What does it mean? Nature. December 13, 2022;**612**: 597-588

[13] ScienceNews. In a breakthrough experiment, nuclear fusion finally makes more energy than it uses. 2024. Available from: https://www.sciencenews.org/article/nuclear-fusion-breakthrough-energy

[14] LLNL—Lawrence Livermore National Laboratory. What Is the National Ignition Facility? 2024b. Available from: https://lasers.llnl.gov/about/what-is-nif

[15] LLNL—Lawrence Livermore National Laboratory. Lawrence Livermore National Laboratory Achieves Fusion Ignition. 2024a. Available from: https://www.llnl.gov/news/lawrence-livermore-national-laboratory-achieves-fusion-ignition

[16] MIT Technology Review. What fusion's breakthrough means for clean

energy. 2024. Available from: https://www.technologyreview.com/2022/12/13/1064898/what-fusions-breakthrough-means-for-clean-energy/

[17] Nuclear Engineering International. Germany announces funding for fusion power plant. Available from: https://www.neimagazine.com/news/newsgermany-announces-funding-for-fusion-power-plant-11612331. [Accessed: March 19, 2024]

[18] WNN—World Nuclear News. Germany aims to build fusion power plant. 2024. Available from: https://www.world-nuclear-news.org/Articles/Germany-aims-to-build-fusion-power-plant. [Accessed: March 14, 2024]